OLD-FASHIONED TOM
page 157

WHITE MANHATTAN
page 84

CROWN HEIGHTS NEGRONI
page 45

TOM O'CONNELL
page 25

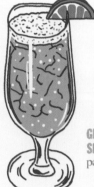

ONE MINT JULEP
page 132

BEHIND THE WALL
page 74

GINGER SHANDY
page 242

VICTORIA'S STONE
page 42

A NIGHT IN TUNISIA
page 268

MAI TAI
page 174

PINEAPPLE RYE CRUSTA
page 68

BROOKLYN EGG CREAM
page 72

GRAVESEND SMASH
page 131

CHIEF GOWANUS MARTINEZ
page 197

OLD MAN DANCE
page 54

PETER THOMAS FORNATALE
CHRIS WERTZ

Brooklyn
SPIRITS

CRAFT DISTILLING ~and COCK-TAILS

FROM THE

World's Hippest Borough

PHOTOGRAPHS BY MAX KELLY

DESIGN BY ERIC SKILLMAN

POWERHOUSE BOOKS pH BROOKLYN, NY

For Susan, whose idea this was,
and for Brooke, whose patience
allowed it to happen.

12.00

CONTENTS

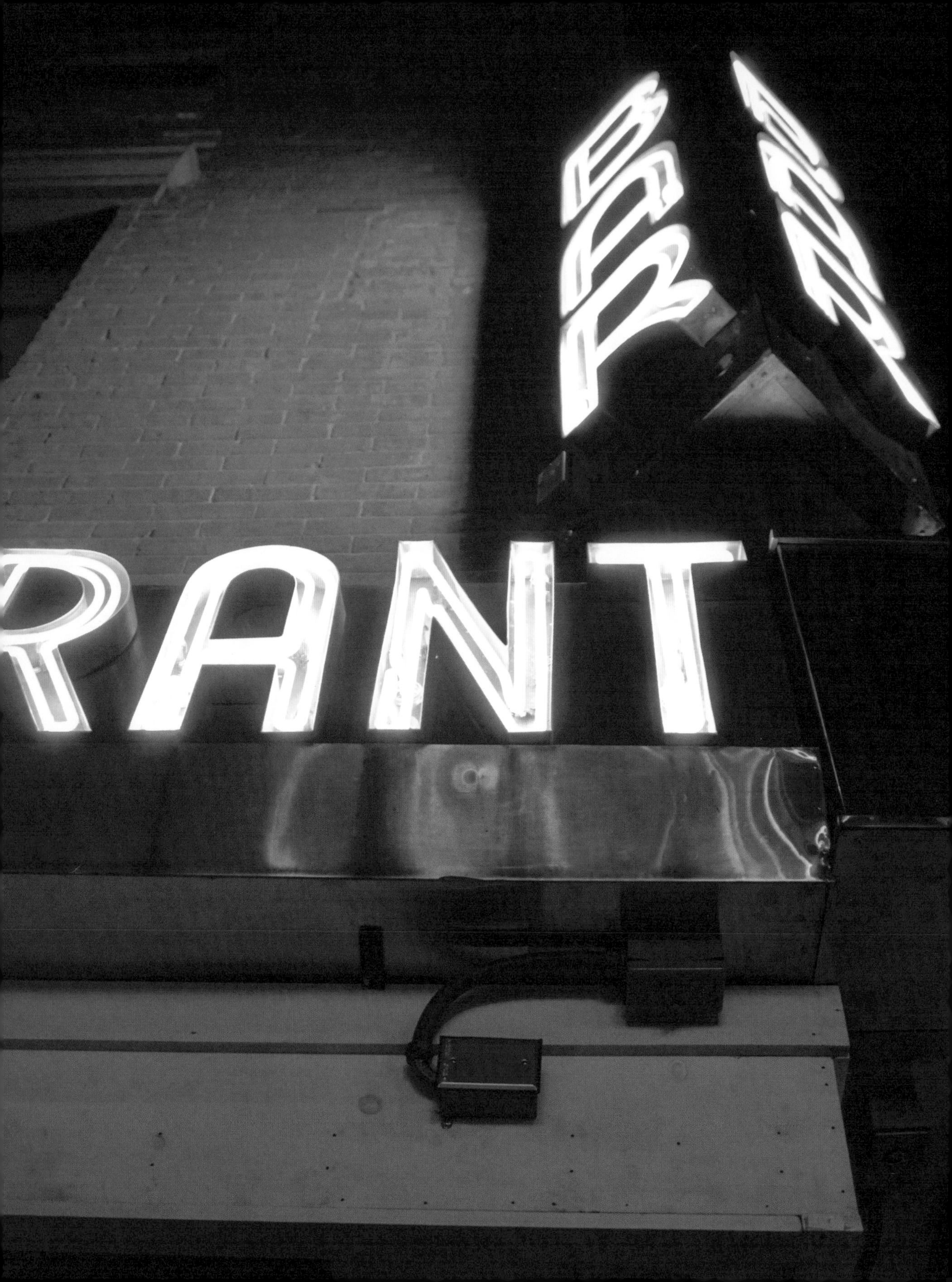
BAR
RANT

I first moved to Brooklyn in 1996, into a tiny apartment with no heat and a rodent problem on 7th Street between 8th and the park. The place was little more than a crash pad, and most of my socializing was done after work in what we then called simply "the city," meaning Manhattan, the place that mattered.

Of course, it wasn't always that way. ▶

The great cocktail historian David Wondrich laid out the facts in a 2013 piece in Esquire: "On January 1, 1898, the city of Brooklyn merged with New York (then confined to Manhattan and the Bronx) and the sparsely populated counties of Queens and Richmond (i.e., Staten Island) to form a new metropolis that would be known as New York City. At the time, Brooklyn was a geographically large and prosperous city of 1.2 million (to Manhattan's 1.8 million) with its own police department, fire department, and city hall; its own downtown; its own theater district; its own clubs and civic organizations, bourgeoisie, history, and destiny. Suddenly, it was a 'borough,' not a city; a part, not a whole. What wasn't merged away began to wither and fade."

I think it's safe to say I arrived in late-fade. Neither the drinking nor dining scenes offered much excitement, at least not in my neighborhood, and it didn't matter that much to me because I didn't have any money anyway. As an assistant in publishing, it wasn't uncommon to spend several nights in a row dining on the same pound of pasta with vegetables, or to drink nothing but the 99-cent 22-ouncers from the corner deli. As for going out, there was just one place I liked. It was called the Park Slope Brewery, and they brewed their own, very good beer. Best of all, one of the bartenders was a WFMU disc jockey, and he played these amazing custom mixtapes (yes, tapes) during his shifts. You could eat a pulled pork sandwich, enjoy a Park Slope Pale Ale, and listen to Michael Shelley expertly start a set with Hank Williams,

end it with the Clash, and have every segue along the way make sense. At some point, the Brewery stopped making its own beer and became just another bar.

Fortunately, everything old is new again. Fifteen years after my arrival, the Brooklyn food-and-drink scene is better than it ever was: farm-to-table restaurants abound, great ethnic spots continue to thrive, and new beer and cocktail bars crop up monthly. And these days, while Brooklyn hasn't officially become its own city, it has done something bigger: it has become synonymous around the world with a lifestyle movement. Wondrich describes a Brooklyn that has "become as much an adjective as it is a place, shorthand for a streetscape abundant in tastefully restored storefronts full of straight-razor barbers, bottlers of unusual pickles, and the like; for a style of urban living that avoids outward swank or gloss—one that's low-rise, low-key, and aggressively local."

We're living in a thrilling time in Brooklyn history. While the rents in Manhattan have exploded, New York's young creative class has fled to Brooklyn, where they are remaking the borough. After decades of being relegated to second-class status, Brooklyn is New York City reclaimed. These days it's not uncommon for me to stay in Brooklyn for weeks at a time. In fact, I think I could stay here for a year and I'd have great places to go out every night. Brooklyn is much more than a place to live. It's a place to live, work, play, eat, and, best of all, to drink.

HOW TO USE THIS BOOK

There was a time not long ago when the idea of whiskey or gin that came from Brooklyn would have been nothing more than a punch line. But in 2002, a change in the law made craft distilling in New York State possible for the first time since before the start of Prohibition. Many producers have taken advantage of this, and their products have quickly gained a foothold in their native borough and beyond. Colin Spoelman, a partner in Kings County Distilling, explains, "The distilling industry is coming back to New York State in a major way. Consumers are sort of tired of the endless branding that comes out of spirits companies. Same stuff, different bottles."

Once upon a time, cocktails were created to mask inferior liquors. Nowadays, the best liquors are sought to pair with inventive infusions to create delicious, original drinks. Because distilling in Brooklyn is in its vigorous youth, a lot of new, exciting spirits are being produced, including several that have garnered national recognition for their quality. One goal of this book is to acquaint readers with the depth and breadth of exciting stuff being made here through profiles of local distillers, infusers, and bitters makers, and even a vermouth producer. We bring you their backstories and their motivations for going into business, and we describe their production processes—particularly in terms of what makes their products stand out among the competition. Each profile is rounded out with cocktail recipes from the producers and from some of the best bartenders and restaurants in Brooklyn. Each chapter also features commentary and exciting recipes from the best local bartenders and restaurants. Another goal of ours is to foster a Do-It-Yourself mentality in our readers, whether you are reading in Kings County or anywhere else around the world.

Of course, many of these craft/DIY ideas associated with food and drink in Brooklyn right now have actually been around forever—putting great local produce to good use, making ingredients yourself, thinking seasonally. When Professor Jerry Thomas wrote his classic Bartender's Guide in 1862, he augmented his cocktail recipes with Do-It-Yourself versions of syrups, bitters, tinctures, and garnishes. The reason? These things weren't commercially available then. Nowadays, while it's easy enough to buy cocktail cherries, syrups, and bitters, you can create really good ingredients less expensively at home. This book will show you how to do just that.

A NOTE ON SOURCES AND SUBSTITUTIONS

A lot of the offerings we describe in these pages are available nearly nationwide—state laws permitting—from online retailers like Astor Wines (astorwines.com). Of course, the best way to buy these bottles is to purchase them at the source. Many of our producers are open for tours and legally able to sell directly to you should you have the chance to visit. We've included the URLs of each of our makers in the book and we encourage you to follow along with us—and ask questions—over on our blog, www.brooklyn-spirits.com.

In the event you can't find one of the spirits called for in a given recipe, you can feel free to be creative and make a substitution. Obviously, we chose what we chose for a reason, but a lot of the fun in making drinks comes from mixing and matching, and we encourage you to do that as well. You should get an idea of what each spirit tastes like from our notes and could sub out accordingly based on online reviews. Again, feel free to contact us and we can try to help you along as well.

A NOTE ON SYRUPS

Where a recipe asks for "simple syrup" or "honey syrup," combine equal parts sugar (or honey) and water in a small sauce pan. Simmer until the sugar is dissolved. Let cool before using. For "rich syrup" use the same process with twice as much sugar (or honey) as water. There will be variations throughout the book that will be described in the appropriate places.

Brad Estabrooke was living what many would consider to be the American Dream. He was working on Wall Street as a trader for Deutsche Bank and living in Brooklyn with his future wife, Liz. The only problem was that Brad *hated* his job. Liz, coincidentally, hated her job, too. Estabrooke recalled, "We were saying to each other, 'Look, these are good jobs that people—many people—would love to have, so why are we so unhappy?'" ▶

They decided that the best approach would be to keep working at their current jobs and save their own money to fund the venture. But in the fall of 2008, the economy tanked, and Estabrooke was laid off.

It was clearly time to put Plan B into effect; only there was no Plan B. Estabrooke had come to a crossroads in his life, but even though he didn't have the money he felt he needed, he couldn't get the distillery idea out of his head. Estabrooke said, "A company was recruiting me for a position for more money than I had ever made, but I couldn't see myself doing it. I really wanted to see what could happen with the distillery idea—fully believing there was no shot in hell that I would ever pull this thing together."

Pull it together he did. Breuckelen Distilling was born in the summer of 2010, through the help of some loans from family members and friends to cover the startup costs. By naming his distillery Breuckelen, spelled the original Dutch way, Estabrooke wanted to show his respect for the once-rich history of distilling in Brooklyn, which had all but disappeared since Prohibition, and also that he represented a continuation of a long revered tradition.

They couldn't help but notice that original, interesting, independent craft businesses were sprouting up all around them in Brooklyn. Estabrooke's lifelong dream was to produce something, anything. Inspired by the independent spirit they saw in their home borough, the couple decided that they wanted in. But what to make?

Estabrooke was aboard an airplane one day. He picked up the in-flight magazine and began to thumb through it. What he saw next caused an epiphany. He said, "The article explained how laws had changed, and for the first time since the start of Prohibition, running a small distillery could be an economically viable business. So, I thought, 'Wow, this is great; too bad Liz would never go for this.' But when I got home and told her about it, she thought it was a great idea."

Estabrooke worked hard to get to where he is now—from unhappy banker to distilling entrepreneur—and he emphasizes the importance of educating one's self. He read whatever materials on distilling he could get his hands on, and he paid close attention to the spirits that he drank. This knowledge was the most valuable thing he carried when he moved into his new space with nothing but a toolbox from his high school days, modest investments from friends and family, and a dream.

It may have felt like a long, slow slog to Estabrooke, but to those on the outside it seems like Breuckelen Distilling has grown remarkably fast. It was a leap of faith and now it looks as if that faith is being rewarded. However, there continue to be challenges, a lack of space being key among them. "We're kind of maxed out on whiskey barrels. In another month, we'll only be able to produce what we're selling. Whatever's empty we can replace, but we need another space." As problems go, that's a good one to have.

As for Estabrooke's perspective on the Brooklyn distilling scene, and his own place within it: "This is great. All these places are open and people are making really great stuff, too. I think it's super-difficult, super capital-intensive, and a ton of work continues to be poured into it. But there is something so great about the people doing what we are doing. We're makers; we are crafting something from scratch. And doing it ourselves, all under one company, one set of people. That, to me, is the part of this business that is so great. We bring in grain, and we send out bottles of spirits. And that, I think, is a really special thing to do."

PROCESS

Estabrooke became animated as he talked about his process, "There are a million ways to make a whiskey. Whiskey is just fermented grain—distilled from grain, and aged in oak. Super simple definition, right? But think about the breadth of varieties of whiskey, and they're all so different and, for the most part, they're all good. So it's really just about how you want to interpret that seemingly simple process. Some of those differences come from what we do, some from the equipment we use, some from the grain itself, or the origin of the grain, some from the oak, some from how long we mature it or even where we mature the barrels, the climate, the building, all these different variables. And learning it all is very much a process of trial and error."

One thing that really sets Breuckelen Distilling apart from the competition is that all of their products are the blended sums of individually crafted parts (more on that in a bit). Breuckelen's grains, all sourced from New York State farms, are milled to specifications in-house (they were the first Brooklyn distillery to do that). Estabrooke and his team don't just throw everything in together and hope for the best; they have a much more nuanced approach.

"We started by making gin. We were always planning to do whiskey, too, but we wanted the gin to taste different. We really loved the character that came from a pure wheat spirit distilled like a whiskey. So we made a wheat spirit as the base for our gin." This was an important decision. Most gins, including the others made in Brooklyn, begin with a sourced neutral base spirit to which botanicals are added before it is distilled again. The decision to use what was essentially a whiskey base he created from scratch brought Estabrooke closer to the tradition of gins distilled in Brooklyn 200 years ago.

Estabrooke continued, "After a couple of months of making gin, we were thinking, 'How are we going to make whiskey?' The wheat spirit tasted great to us, just coming off the still, so we decided to put it in oak and see how it came out." Thus was born the 77 Wheat Whiskey, a simple, elegant expression of a local spirit, distilled from 100 percent local wheat and aged in new American oak barrels.

And the rye? That was sort of a happy accident, too. Estabrooke said, "We had enough grain to do a mash of 90 percent rye and 10 percent corn, so we filled another barrel with that. That was the only grain in the distillery that day and so I figured, 'Why not?'"

"For the 77 Rye and Corn we now do a rye mash and rye distillation of 100 percent rye, then we do the same for the corn, then we blend them. We do the same thing with the gin; we distill each botanical individually because it gives us the ability to get the best extract of juniper, and then the best extract of lemon, and then blend those, rather than get the best extract of juniper and lemon together which is not going to be the same."

Estabrooke exhibited great humility as he said this, but what he is talking about is a really novel approach, and the results speak for themselves.

SHOULD ONE MIX CRAFT WHISKEYS?

Throughout the project, we've encountered many people, distillers and bartenders alike, who are excited about the idea of making drinks with craft whiskeys. But there's also been plenty of resistance. More than once we've heard a variation on the line, "I/They worked hard to make it a certain way. Why would you want to change how it tastes?"

It's a valid point. Colin Spoelman of Kings County Distilling has an interesting perspective, "I think it's just really a question of authorship, and who is responsible for what you are drinking. Craft distilleries want it to be them, and cocktail makers want it to be them, and along the way the narrative gets a little muddled when you have two authors."

Some of it comes down to simple economics. Small-scale distillers are looking for consumers to taste what's special about their products so they can grow. They can't compete on price, so they need to be perceived as special. Legendary barman and one-time Brooklyn resident Dale DeGroff related a funny story about the way cognac producers' perspective about their own product has changed over time because of economics. "Back in the 70s, cognac guys used to say, 'Don't you dare put my beautiful spirit in a cocktail. You'll ruin it.' Now they're falling all over themselves just to get it in cocktails—they're happy to serve it as a chilled shot!"

Typical bar owners aren't looking to use spirits that are relatively expensive to buy in cocktails where something solid but less expensive will do. We get both perspectives, for sure.

But we're happy to report that some restaurants and bars are giving away a bit of their margin to support the local spirits scene. You'll meet a lot of them in this book. And in sharing their cocktail recipes, we hope to also honor the work of the spirit producers. We don't have any interest in dumping a lot of sugar in carefully crafted spirits, masking the very essence of what they are. And while we both enjoy that many of the spirits featured in this book can be consumed straight or with a little ice—and even recommend that they be served this way—we have also found that certain cocktails can actually accentuate what it is we like about the base spirits in the first place. It is this type of drink—call them spirit-forward cocktails—that we'll be featuring in this book.

Nate Dumas from New York Distilling Company, who has a dual background in distilling and mixology, seemed like the perfect guy to ask about this. "I have run into people, some of them distillers, who feel that a quality spirit is somehow sullied by being used in a mixed drink," he said. "Maybe they're not wrong in some cases—there are plenty of lousy cocktails out there. But I think that idea is outdated. You can't come to the Shanty, or go to places like Clover Club, or any of the other professional cocktail bars that have popped up all over the country, and not feel that the spirits are being celebrated in the drinks."

LE SUBTIL

2 1/2 oz Glorious Gin
1/2 oz sweet vermouth
1/2 oz Cardamaro
1 tsp bergamot cordial
(*or 3 dashes Angostura orange bitters*)
grapefruit twist

BERGAMOT CORDIAL

Use the same basic recipe as the lime cordial you'll find on page 191. Finding bergamots won't be easy. But if you manage it, simply replace the limes with the equivalent amount of bergamots by weight (approximately 2000 grams or 4.6 lbs) and follow the instructions. In the likely event that you can't find bergamots, you can substitute a mix of two grapefruit and nine limes. Or, as suggested above, use Angostura orange bitters.

This is a contribution from Toby Cecchini of Long Island Bar. As he describes it, "This was something I made at home a few years ago for myself, finishing it with a tincture I'd made of pomelo and sour Seville orange peels, wanting something very subtle to play off of Bols Genever." It first premiered on the cocktail menu of the famed cocktail bar Death & Co. When Toby tried Glorious Gin, he decided to tinker with this recipe again. He said, "The genever-like aspect of Brad's gin prompted me to incorporate that into the drink. I was recently able to acquire fresh bergamot citrus for the first time, so I made a cordial with that and used it to finish the drink: it has a remarkable punch. This is my favorite version yet."

LONG ISLAND BAR
(Cobble Hill)

Many of the best and brightest in the Brooklyn cocktail scene are refugees from Manhattan who came to Brooklyn for bigger space, more freedom, and other benefits of the borough. Their great talents add to an emerging scene of enthusiastic newcomers and hard workers.

Toby Cecchini, innovator behind the Cosmopolitan—and thereby an instigator of the 90s cocktail craze that brought sex to the city—has recently taken up shop in the decidedly less cosmopolitan borough of Brooklyn. It's important to note that Toby's other contributions to the world of cocktails dwarf the Cosmopolitan: he is a master of subtle flavors behind the bar and a brilliant writer and historian, and his pioneering work in the field of Do-It-Yourself mixology pretty much means this book wouldn't exist without him.

In October 2013, he and partner Joel Tompkins reopened and revamped the legendary and long-shuttered Long Island Restaurant and Bar, a former haunt of longshoremen on Atlantic Avenue in Cobble Hill. In accordance with the former owner's wishes, and in deference to their own great taste, they largely left the design features of the Spanish diner intact. But in place of the Spanish fare, world class cocktails now command the bar space, attracting discerning neighborhood tipplers along with luminaries from the New York City bar world like Dale DeGroff and David Wondrich. In fact, Wondrich named Long Island Bar one of the top 25 bars in the United States in a 2014 *Esquire* piece. Some Brooklyn cocktail bars are local bars, some are destination bars—Long Island Bar is both.

TOM O'CONNELL

1 1/2 oz Glorious Gin
1 tbsp fresh squeezed lime juice
1 tbsp simple syrup (*see page 11*)
soda water
5 fresh thyme leaves

*Combine first three ingredients and mix
well. Pour over ice, top with soda water,
and garnish with thyme leaves.*

Every wedding deserves a good cocktail. In the summer of 2012, Brad married the love of his life, Liz O'Connell, and this was the drink they served at their reception in Maine. It's a smart play on the classic Tom Collins where the thyme adds a savory element and makes the drink a lot of fun to look at.

ONE MORE SIP

2 oz 77 Local Rye and Corn
1 1/2 oz Blueberry-Honey Syrup
1/2 oz lemon juice
Splash Kübler Absinthe
3 drops BHB Meyer lemon bitters
lemon wheel
3 blueberries

*Put the first three ingredients in a cocktail shaker filled most
of the way with ice. Shake, then strain over fresh ice in a
rocks glass rinsed with Kübler Absinthe. Top with the BHB
Meyer lemon bitters, a lemon wheel, and the blueberries.*

BLUEBERRY-HONEY SYRUP

1 cup water
1 cup honey
1 cup fresh blueberries

*Combine the water and sugar over medium heat and mix
in the blueberries, crushing them a bit with the back of a
spoon. When the mixture just gets to a boil, turn off the
heat, stir to combine, and let sit until cool. Strain through a
sieve into a bottle or jar with a tight-fitting lid.*

Mark Buettler of Brooklyn Hemispherical Bitters has a background working in restaurants, and that shows in his careful selection of main ingredients in his bitters, each of which looks to clearly express a single flavor. His Meyer lemon bitters are among our favorites. Pete likes them so much that he's been known to sneak a bottle around with him at Saratoga Racecourse and use them to doctor various cocktails on the fly.

This recipe is actually on the bottle for the BHB Meyer lemon bitters as a bourbon drink, but we like the way it combines with the 77 Local Rye and Corn whiskey and Blueberry-Honey Syrup to create a fun, summery take on a whiskey sour. In a cocktail like this, with so many strong flavors, the 77 Local Rye and Corn still manages to keep distinct whiskey flavor at the center of the drink.

SICILIAN OLD-FASHIONED

2 oz 77 Local Rye and Corn

1 oz Amaro Segesta*

3/4 oz rich syrup (*preferably a raw sugar like Demerara; see page 11*)

1 orange slice (*blood orange slice if available*)

1 good quality or homemade cocktail cherry (*see page 159*)

2 dashes aromatic bitters

Muddle the fruit, bitters, and sugar in the bottom of a rocks glass. Add ice, amaro, and whiskey, then stir.

*Averna, a more common Sicilian amaro, would be a decent substitute, but the Segesta, which Danny calls an "old man Sicilian amaro" is worth seeking out.

Want to start a fight between two Brooklyn bartenders? Bring up the Old-Fashioned. It's a drink with separate identities that aren't easy to reconcile. Most of us grew up, like Danny Kent, the bar manager of Locanda Vini e Olii, knowing the Old-Fashioned as a sweet drink preferred by various older relatives. It consisted of muddled fruit (cherry and orange), sugar, bitters, and whiskey or brandy (in Wisconsin anyway).

This version, however, is not the historically accurate version. The proper Old-Fashioned (Old Old-Fashioned? New Old-Fashioned?) was a simpler concoction, omitting that muddled fruit that was such a defining feature of the 70s and 80s version we all grew up watching our relatives drink.

The proponents of the proper Old-Fashioned typically turn their noses up at the fruited version—probably out of antipathy to the bright-red, toxic-looking cherry that so frequently resided within. We have some sympathy for this point of view. We're certainly not interested in eating a cherry that looks likely to last longer than Keith Richards and the cockroaches. But that said, this is a solved problem. Homemade cocktail cherries can be very good (see page 159), plus there are several brands of decent cherries now commercially available.

In our travels, we've seen a few nods to the muddled fruit Old-Fashioneds of our youth. The slyest version came from Alla Lapuschik at Post Office. She simply makes the old-school Old-Fashioned and drops a Luxardo cocktail cherry in the thing. Henry Lopez at Lucey's Lounge does an excellent, crafty version of the muddled Old-Fashioned. But the one we're going to share for the book comes from our man Danny Kent. This drink is his take on the whole debate.

THE WEDDING SLINGER

1 1/4 oz 77 Local Rye and Corn
1/2 oz Carpano Antica Formula sweet vermouth
2 dashes orange bitters
ginger ale

Garnish with at least one good quality or homemade cocktail cherry (see page 159).

Place all ingredients besides the garnish in a shaker mostly filled with ice. Shake and strain into a collins glass filled with ice and top with ginger ale (preferably from syrup made by Pickett Brothers or Brooklyn's Morris Kitchen).

The Wedding Slinger was born out of conflicting needs: the need for numbness and the need for temperance. Chris was employed as a bartender at the Blind Tiger Ale House in downtown Manhattan on September 11, 2001. His clientele that week included first responders and soot-covered camera crews. He attended a wedding on the coast of Maine on September 14, 2001. The long, slow drive up the coast of New England was peaceful, but left him too much time to think. The first thing Chris reached for when he got to the wedding venue was an old favorite: a Manhattan cocktail. The first one went down like a shot. Second one, same thing. He decided a wedding was no place to have a booze-fueled breakdown, so Chris asked the bartender to top off his third Manhattan with some ginger ale. Yummy. He asked for another, this time with a little less whiskey and a little more ginger ale. Yummier.

When he got back to The Blind Tiger later that week, he fooled with the recipe, replacing the sweet bourbon from the wedding with a drier rye whiskey. The drink developed laboriously through the thoughtful trial of his fair patrons. What they wouldn't do for a free drink.

FRANNY'S

(Park Slope)

When Franny's opened on Flatbush Avenue in 2004, it represented a sea change in the Brooklyn restaurant world. An authentic pizzeria with a seasonal menu, great cocktails, and an eclectic wine list, Franny's soon became one of the big boys on the scene, compelling the owners to move to a larger space down the block in 2013. Our author Pete and his wife, Susan, are among the legion of Franny's regulars who feel lucky to have the place so nearby.

Francine Stephens and her husband Andrew Feinberg have always embraced a creative approach to cocktails and have recently embraced the local spirits scene. Stephens described the evolution of the Franny's cocktail: "Our entire focus at Franny's has always been local and seasonal, with both the food and the cocktails. When we opened in 2004, the cocktails were less about the spirits than they were about the accompaniments—rhubarb juice and ramps in the spring, greenmarket vegetables in the summer. But 10 years ago, of course, all these local spirits did not exist. It just seems like the perfect evolution that now we have spirit offerings that are local and made by hand by people who really care about making beautiful spirits. In terms of who we are, it makes perfect sense. And now that we've opened our new restaurant Marco's, it was the perfect timing for us to open a bar with almost exclusively local spirits."

Having dined at Franny's extensively over the past decade, Pete knew that there had to be more than just being local to get Francine so excited about the Brooklyn spirit-makers. "They're new, they're different, they're interesting," said Stephens. "The local spirits help us engage our customers. People ask questions about them, so there's inherent education going on. And best of all, when people get their drinks, they're loving them."

BROOKLYN COCKTAIL

1 1/2 oz 77 Wheat Whiskey
3/4 oz Martini and Rossi
sweet vermouth
1/2 oz fresh lemon juice
1/4 oz simple syrup (*see page 11*)
lemon twist

*In a cocktail shaker filled with
ice, add all the ingredients
except the garnish. Shake hard
and strain into a cocktail glass,
preferably chilled. Express and
drop the twist.*

Francine Stephens knew right away the drink she wanted to put at the center of her cocktail program.

"There were a handful of cocktails that I designed myself. I was opening a restaurant in Brooklyn and I didn't really know what I was doing, but it was Brooklyn, not Manhattan, and I wanted to take a stance. At that time, the Brooklyn was really in obscurity as a cocktail and I wanted it to have a resurgence. That was my goal. And it's still on the menu after 10 years. It is, hands down, our most popular cocktail."

We think if you taste this smart reboot of the Brooklyn, you'll understand why. It's basically a Manhattan sour, and the 77 Wheat Whiskey gives it a particularly lovely, spicy finish. "We always made the drink with Maker's Mark and I was very hesitant to change that up, but we have used a few different local and/or craft spirits, and the drink still works in many cases. The 77 Wheat Whiskey captures the essence of what our Brooklyn is all about."

SUBURBAN

2 oz 77 Local Rye and Corn
3/4 oz dark rum
3/4 oz ruby port
dash orange bitters
dash Angostura aromatic bitters

*Place all ingredients in a cocktail
shaker mostly filled with ice. Shake
and strain into a rocks glass with
fresh ice. Garnish with a candied
cherry, brandied cherry, an
orange twist, or some combination
of the above (as in the photo).*

This cocktail, the first of many racetrack themed drinks in this book, was suggested by Brad Estabrooke, one of his favorite ways to feature his 77 Rye and Corn Whiskey.

For the record, the Suburban is a famed stakes race first run at Sheepshead Bay in Brooklyn in 1884, and is still run today out at Belmont Park, on the border of Queens and Long Island. It's been won by an impressive array of horses including Imp, Assault, Tom Fool, Kelso (twice), Forego, Easy Goer, Skip Away, and many, many more. The 2009 winner was a horse named Dry Martini, which is rather ironic given that a dry Martini is actually about the farthest away one could get from this highly unusual drink—the only libation we can name that thinks to combine whiskey, port, and rum.

OAK-AGED NEGRONI

1 oz Glorious Gin: Oaked
1 oz Campari
1 oz Cocchi Vermouth di Torino

Place all ingredients in a shaker that is mostly filled with ice. Stir, then strain into a rocks glass with ice or a chilled cocktail glass up. Garnish with an orange slice.

This variation on a classic Negroni replaces dry gin with an oak-aged version of the Glorious Gin. In the process, it reinvents the drink, taking it to a halfway point between the Negroni you know and love and a Boulevardier (which is a Negroni with whiskey instead of gin).

One of the many cool things Danny Kent has done in his time managing the bar program at Locanda Vini e Olii is to devote an entire section of his cocktail menu to various takes on the Negroni. This drink is a worthy addition to that list.

This is your typical story about a former underwear model leaving a career in marketing after a serious cancer scare to pursue his dream of bringing to market a product that nobody had ever heard of and somehow convincing people to buy and use it. Sound impossible? Perhaps. But Jack Summers thrives on the impossible. ▶

In 2012, Jack Summers formed the company Jack From Brooklyn to produce and distribute a new liqueur called Sorel, which was based on a traditional beverage that Jack had been making in his home for more than 15 years.

In the creation of Sorel, Jack reached deep into the tradition his grandparents brought with them from Barbados, and the flavors that surrounded him in New York as a child of West Indian ancestry, to devise the ultimate homage to his ancestors: the flavor of the Caribbean in a bottle. As Jack deftly describes it, "If mom's apple pie and dad's liquor cabinet got together and got busy and you put it in a bottle, you get Sorel."

Explained Jack, "My grandparents emigrated from Barbados [to New York City] in the 1920s, and like other immigrants, they prepared ethnic foods to remind themselves of home." Through his grandparents, Jack grew up with versions of a Bajan hibiscus tea called "sorrel" (two Rs), in his home. Sorrel tea is an herbal drink from the West Indies, made from dried hibiscus petals and blended with spices and roots like cinnamon and ginger, and sometimes (for special occasions) spiked with rum. This traditional homebrew has been a staple all over the West Indies for centuries. No two recipes are identical, and families pride themselves on their individual expressions of the summer cooler.

Jack had been making homemade sorrel tea in his kitchen for more than a decade when he was diagnosed with what he was told was fatal cancer, in 2010. It turned out that the tumor was benign, and Jack made a full recovery. Moreover, after his brush with death, he

vowed to utilize his new lease on life to the fullest. Encouraged by the appreciation he had received from his friends for his homemade sorrel tea, Jack put the corporate world behind him and began a new career structured around something personal and meaningful. Jack said, "I turned my back on 25 years in corporate America to put my heritage in a bottle." He tinkered with the recipe and developed a commercial process, eventually debuting his personal brand of island-spiked hibiscus tea, which he dubbed Sorel.

Sorel is bursting with spices, which are shipped to his factory in Red Hook and blended there to make a unique, tannic, satisfying liqueur that's both potable and educational. It's a huge risk to sell something that most people have never seen before. But in Brooklyn, with its very large West Indian population and omnipresent craft-cocktail culture, it seems that Jack is giving the community something they've been waiting for—even if they don't know it yet.

Sorel was an almost immediate hit among local tipplers and aficionados alike. Wine Enthusiast awarded Sorel 91 points in the autumn of 2012, calling the liqueur "addictively sippable." Just a couple of months later, the *Spirit Journal*, an industry standard on alcoholic beverages published by the eminent F. Paul Pacult, gave Sorel five stars, its highest honor. The review called it "Perfectly rendered in terms of spice, floral aspect, acid freshness, and fruitiness. All the flavors are pulling the wagon in unison. Love it. Unlimited cocktail potential."

Since its relaunch after Hurricane Sandy in January 2013 [see sidebar], Sorel is now distributed in more than 20 states from Maine to Oregon, and from Minnesota to Louisiana.

"I turned my back on 25 years in corporate America to put my heritage in a bottle."
—JACK SUMMERS

It can also be found in Canada, Australia, and one place where it should feel right at home— Trinidad and Tobago.

What's next for Jack? Right now he's focusing on building brand awareness domestically and abroad. But he does have his eyes on some new products in the not so distant future. He told us, "Here's the theory behind Jack From Brooklyn: Sorel was a regional, niche beverage you only knew about or had access to if you had Caribbean ancestry. I believe there are many other beverages out there, waiting to be 'discovered.' Find these beverages, solve their logistical obstacles, and market them in a way that makes them accessible."

In making Sorel, Jack starts with high quality spices like Moroccan hibiscus, Nigerian ginger, Indonesian cassia and nutmeg, and Brazilian clove. He extracts their flavors in water in what he calls "a giant tea kettle." This is similar to the sorrel tea found throughout the Caribbean. Once the essential oils of the spices are released into the water, Jack makes an adult beverage out of the mix by blending in 100% organic kosher neutral wheat grain alcohol. Lastly, he uses pure cane sugar to add some sweetness and neutralize the acidity of the hibiscus.

HURRICANE SANDY

Things were looking up for Jack Summers until Hurricane Sandy ravaged Red Hook in October 2012. Starting a small business is always fraught with unforeseen difficulties, but nothing could have prepared Jack for Sandy. Well-fortified with superlatives from industry experts, and just five months after Sorel first hit the shelves, Jack lost it all: "Everything was destroyed—commodities, product, electrical equipment—and the 150-year-old building that houses the distillery sustained major infrastructure damage."

Jack's Red Hook workshop was under five feet of water as a result of flooding—a devastating blow that could have easily ended the promising future of the brand new business and the realization of a lifelong dream. The "superstorm," which washed away homes and destroyed entire towns, left the waterfront neighborhood of Red Hook inundated and facing a seemingly impossible recovery.

Another business that was nearly washed away by Sandy was Dry Dock Wine and Spirits. "If any of us down here doubted the nobility that lies within people, you couldn't walk away from Sandy not feeling better about being a human. The way that people showed up to help was remarkable," owner Mary Dudine Kyle told us. "Everything you see here now was gone. Our floors had buckled and there was four feet of water. But we were able to reopen in a space down the block in four days thanks to everybody's help. It was an incredible experience."

Similarly, Jack received help from the community, including support from local spirit-makers like Josh Morton from Barrow's Intense and Allen Katz of New York Distilling Company.

In the end, it was going to take more than a major weather event to derail Jack and his plans. Two years later, Jack's Sorel stands out as a testament to post-storm possibility across the region.

MIXED BAG

4 oz chilled Champagne
1/4 oz Sorel
1 sugar cube
dash aromatic bitters

Build in a Champagne flute
or cocktail coupe.

Note on Champagne:
We used Champagne "Prestige" Brut, Duc de
Romet NV, a great value for Champagne.
Feel free to use more or less Sorel to your taste.

When Pete's dad was searching for a name for his new radio show—an eclectic mix of folk, rock, jazz, spoken word, comedy, and whatever he felt like playing—he chose the name Mixed Bag. It was a good, literal description but also an homage to one of his favorite artists, the great Richie Havens, a Brooklyn native with a singular voice whose music, particularly his interpretation of other artists' songs, will live forever. Havens' *Mixed Bag* album remains a favorite of both of ours, and the copy he signed to Pete's dad, "a friend forever," as he put it, is one of Pete's most prized possessions.

One of Richie's many achievements was opening the Woodstock festival back in 1969—though only after the originally slated opening act, Tim Hardin, declined to appear (one account had him hiding under the stage in fear at the prospect of performing in front of 500,000 people).

Havens was undeterred. He went on and because no else was ready to play, he just kept playing—and playing and playing and playing. His epic set was inspiring, especially his improvised anthem that's come to be known simply as "Freedom." It set the tone for the whole weekend, and as time went on, Havens became the living embodiment of the Woodstock ethos—a loving, charitable man and an incredible musician.

The beverage of choice backstage at Woodstock, at least according to Brooklyn native Arlo Guthrie, was Champagne. The Mixed Bag offers up a classic Champagne cocktail, with the addition of a splash of Sorel liqueur, a nod to Richie Havens' West Indian heritage on his mother's side.

VICTORIA'S STONE

1 1/2 oz Dorthy Parker American Gin

1 oz Sorel

1/4 fresh plum, thinly sliced

1/2 oz rich honey syrup (*see page 11*)

1/4 oz lemon juice

Put all the ingredients in a cocktail shaker with ice and shake vigorously. Pour contents into a rocks glass.

Ditmas Park has become one of the cooler areas in Brooklyn to eat and drink, with places like The Farm on Adderley and The Sycamore leading the way. The Castello Plan is one of the later (and better) additions to the scene out there on Cortelyou Road. Named for Jacques Cortelyou's first map of Manhattan, it's a charming spot known more for wine than cocktails, but the place sports a cool cocktail menu featuring a couple of local spirit-makers.

Gabrielle Delaney of The Castello Plan designed this drink. We asked her where she got her inspiration: "I like the Christmas-y spice notes in the Sorel, and that got me thinking about Christmas, which is where the idea to pair the Sorel with plums came from, and in early fall we were getting these really great plums, so it just made sense. The Sorel itself acts as a sweetener, so I thought that adding simple syrup would make it too sweet, but I liked the idea of adding a little honey. It adds a nice viscosity and an overall roundness to the drink."

Sorel is unlike any other spirit on the market. The color, which is drawn from the hibiscus petals, is a stunningly deep magenta—a perfect hue to brighten any cocktail. The flavors of island spices—cloves, nutmeg, ginger, and cinnamon—all marry well with the plum in this drink.

> "I think Brooklyn has developed a spirit of, 'OK, we're going to do it ourselves.' We're going to make it by hand; we're going to be artisans; we're going to be actual people who make things we need and want."
>
> —DEL PEDRO

CROWN HEIGHTS NEGRONI

2 oz Tanqueray Malacca Gin

1/2 oz Sorel

1/2 oz Carpano Antica Formula sweet vermouth

1/2 oz Campari

Add ingredients to a pre-chilled mixing glass. Stir in a shaker or pitcher mostly filled with ice, then strain into a cocktail glass. Garnish with a dehydrated orange slice.

FOR THE DEHYDRATED ORANGE SLICE:

Cut mandarin oranges into thin to medium-thick slices. Place the orange slices in a dehydrator (Del uses a small Nesco four-shelf "American Harvest" unit), set to 140 degrees, and leave for 12–15 hours. Leave the slices to cool and set for about 30 minutes before removing them from the dehydrator. Store in a bag that can be sealed or airtight container and keep refrigerated.

To Chris, the name "Del Pedro" was once like a curse. He was bartending at the Blind Tiger Ale House in the West Village where he barely ever made a cocktail more complicated than a Vodka and Tonic, when suddenly more and more of his regulars began asking for what were then exotic drinks, like Negronis. He asked them where they were getting their cocktails, and they all answered, "Del makes them at Grange Hall around the corner." Damn that Del. The cat was out of the bag. From then on, littered among an infinite number of impossible beer questions was, "What kind of sweet vermouth do you use?" Del was an important, early pioneer of the most recent incarnation of the cocktail movement. His influence over drinkers and bartenders alike while at the celebrated Grange Hall is still being felt across New York City. (That Chris is now writing about Negronis should be proof of that!) Add to his list of accolades that his new place, Tooker Alley, in Crown Heights, is a favorite Brooklyn bar of cocktail guru David Wondrich. It's no wonder that Tooker Alley is popular: Del brought the Negroni with him.

TOOKER ALLEY

(Crown Heights)

Del Pedro of Tooker Alley knows why Brooklyn has become such an epicenter for craft culture: "When I moved to Brooklyn, I felt like I was once again surrounded by culturally focused and astute people. The city had gotten to a point where it was 'Get Rich or Get Out' basically. In Brooklyn, it didn't feel like money ran everything; it felt like there were other things that were driving the decisions people made. Brooklyn, to me, still felt like it had elements of 70s, 80s, and early 90s New York, when it was more culturally driven, rather than purely consumption-driven. I have an IWW T-shirt, and right after I moved to Brooklyn, I'm standing on a street corner in Bed-Stuy, and somebody riding by on a bicycle goes, 'Great T-Shirt, bro!' And I'm thinking, 'That's not going to happen in Manhattan.' They'd be hostile or they wouldn't know what it was.

"I particularly fell in love with Bed-Stuy, Clinton Hill, Fort Greene—the whole extended area. It was a revelation. I felt like I belonged in the city again. And I think Brooklyn has developed a spirit of, 'OK, we're going to do it ourselves.' We're going to make it by hand; we're going to be artisans; we're going to be actual people who make things we need and want. And yeah, we'll make less money, and it'll take more time, but it's going to be so much more gratifying. Of course this cocktail thing fits right in with that."

CAMPFIRE SONG

1 1/2 oz Sorel

1 1/2 oz Laird's Apple Brandy

1 oz freshly squeezed lemon juice

3/4 oz smoked maple syrup

Combine ingredients, shake with ice, and fine strain into a cinnamon/apple/sugar rimmed Martini glass.

CINNAMON SUGAR RIM:

1 oz ground apple chips (*Use any baked—not fried!—variety and pulverize in a small coffee grinder until powder.*)

1 oz ground cinnamon

2 oz granulated white sugar

1/4 oz ground cloves

1 pinch cayenne pepper

A NOTE ON THE SMOKED MAPLE SYRUP:

The syrup in the recipe is from a company in Vermont (sugarbobsfinestkind.com) and must be diluted with regular maple syrup, as it is much too strong to use straight from the bottle. Bruce's ratio is one part smoked syrup to five parts regular syrup.

Like Batman, this cocktail has a secret identity. In the winter-time, it is known as it is in this book, "Campfire Song." But during the warmer months, it's been known to shift glasses and garnishes and embark on a journey over ice, transforming into a long, tall sipper known as a Kumbaya Cooler. The connection between the two names should be obvious.

Designed by Bruce Ramsay, formerly of The Huckleberry Bar, the drink shines in either iteration. The mulling spices of the Sorel are a natural match for the apple brandy, and the tartness of the hibiscus petals works perfectly against the smoked maple syrup. The combination of savory, sweet, bitter and sour makes us think of a good barbecue sauce.

(To convert the Campfire Song to a Kumbaya Cooler, use a collins glass filled with ice instead of a cocktail glass and garnish with Granny Smith apple slices. A straw would be nice.)

HUCKLEBERRY BAR

(East Williamsburg)

It makes sense that Huckleberry Bar would be good at the business of hospitality—they learned from one of the best. Co-owners Stephanie Schneider and Andrew Boggs worked for Danny Meyer and the Union Square Hospitality Group for many years before venturing out on their own, and you can tell the place was opened by people with experience.

Huckleberry Bar in East Williamsburg is a place where you can get a great drink in a relaxed atmosphere. That idea speaks to the core of who Schneider and Boggs are. Their motto is "fancy without the fussy" and they certainly live up to that.

The attention to detail shows in the very setup of the bar—all super organized. Even the entertainment calendar at Huckleberry Bar is well-organized: Monday is movie night, local progressive DJs take over on Tuesdays, Wednesday is for wine lovers, and the weekends are... well, the weekends are the weekends—crowded and young, but still fun.

MAE WEST INDIAN

2 oz Owney's rum
1 oz Sorel
3/4 oz lime juice
lime wheel
candied hibiscus flower*

*Build in shaker, with ice. Shake and strain over ice** in a collins glass. Using a skewer, affix candied hibiscus to lime wheel and garnish.*

*Candied hibiscus is available in certain gourmet stores, stores that specialize in West Indian produce, or via wildhibiscus.com.

**For this cocktail, Alicia used a cube made from a long collins cube mold from Cocktail Kingdom (cocktailkingdom.com).

All the bartenders in this book have had people who have influenced them. Dale DeGroff worked with the legendary Joe Baum at the Rainbow Room. Julie Reiner and Adam Kolesar were heavily influenced by Dale DeGroff. That's just the way things go when it comes to making great drinks. Alicia Blegen of Marietta has the most unusual "mentor" of any of our bartenders. She's been heavily influenced by Bushwick native Mae West, who was a pioneering actress, singer, comedienne, and sex bomb—both in vaudeville and in Hollywood. Blegen explained how West has inspired her: "I've always been impressed by her flair, chutzpah, and brassiness. She wasn't afraid to kick your ass or to go toe-to-toe with you. And she had this ownership over her body that was many decades ahead of its time."

This cocktail, designed by Blegen in honor of Mae West, and the West Indian population near Marietta, is a fine, boozy tribute to both.

THE FISCAL CLIFF

2 oz Old Overholt rye
1/2 oz Sorel
1/3 oz St. Germain
splash of prosecco (*McDermott often uses crémant to the same effect*)
orange twist

Place the first four ingredients in a cocktail shaker mostly filled with ice. Shake enthusiastically. Strain into a chilled cocktail glass and float the prosecco on top. Rub the rim with an orange peel and place in the drink.

This drink was born out of necessity, Good Fork bartender Matty McDermott explained to us. It debuted on New Year's Eve 2012, which also happened to be The Good Fork's re-opening night after the restaurant nearly got wiped out by Hurricane Sandy. "We were looking at our stocks and we didn't have much," recalled McDermott. "After two and a half months of being closed, we didn't have a lot of money. So the owner, Ben Schneider, had the idea to come up with a cocktail called The Fiscal Cliff." Schneider charged Matty with coming up with an actual drink to match the name. He surveyed what was left and built this drink around Old Overholt rye and Sorel. The resulting concoction is a fun combination of elements all brought together by the sparkling effervescence of the prosecco float at the end.

The drink was a hit that night and remains popular to this day.

OTB

(Williamsburg)

The sad result of poor planning, political mismanagement, and simple technological progress, New York City's OTB (Off Track Betting) parlors, where one could bet horses from seemingly any neighborhood in the city, are no longer. Fortunately, we now have this bar instead—and these days, with a decent Wi-Fi connection, you can make a bet right from here, though you may face the harsh judgment of peers for being a degenerate.

Big horse racing fan that he is, Pete has spent a lot of time in his life at various OTBs, but he'd never seen an OTB like this tricked-out neighborhood bar in Williamsburg, across the street from Peter Luger. Like its sister establishment Post Office, OTB's name is Charles Bukowski-inspired. Though it's hard to imagine Bukowski himself snacking on escargot (a signature appetizer here), it is easy enough to imagine him downing a few shots and digging the sexy, retro art on the walls.

OLD MAN DANCE

1 oz rye whiskey
1/2 oz Sorel
1/2 oz Ramazzotti

Stir in a cocktail shaker full of ice and strain into a coupe or cocktail glass. Garnish with two orange slices.

This whiskey cocktail comes from Post Office in Williamsburg. Post Office's Bukowski-inspired name and cool atmosphere make it one of the more fun whiskey bars in Brooklyn. While the wall behind the bar suggests you might be best off ordering your spirit neat or perhaps with a rock or two, you can also get good, spirit-focused cocktails. This isn't a surprise, since one of the owners, Alla Lapushchik, helped open the renowned cocktail bar Death and Company in Manhattan, in what now must seem like another life.

Alla has worked with Kai Parrott-Wolfe for many years, often sharing shifts behind the bar at Post Office and OTB. When they work together, they like to dance. Or more accurately, Alla likes to dance. On the rare occasions when Kai decides to channel his inner Deney Terrio, Alla makes fun of him for dancing like an old man. And thus a drink name was born.

What's cool about this drink—other than the name, and fun, double-wheel garnish—is that it shows how Sorel, with its winter spice notes, can combine nicely with brown spirits as well as clear ones.

You don't have to be a rocket scientist to succeed on the Brooklyn distilling scene, but it certainly doesn't hurt. And Dan Preston is literally a rocket scientist. So perhaps it should come as no surprise that Cacao Prieto has hit the ground running. ▶

In his office above the distillery one workday, Preston came off as an extremely driven individual: intelligent, knowledgeable, and creative. Think Willy Wonka as a college professor and you're in the ballpark.

What exactly did he do for a living before distilling? "We ran a defense company for many years," he explained. "We built really high-tech toys for the military."

Preston did more than run the company; he also owned it. The turning point came the day his company received a very large contract. Fulfilling the contact would have meant bloating the company to around 500 employees from 25, and Preston wasn't into the idea, so he decided to sell the company instead. "I had to stay for a year under an employment contract, but during that period I went from doing 100-hour workweeks to doing 40-hour weeks. I needed a diversion. A friend of mine was a mixologist for Sasha Petraske [owner of the famed cocktail bar Milk and Honey], and my friend talked me into building a bar. Now it seems like utter lunacy, but at the time I thought it sounded fun."

The bar is called Lingua Botanica, which is Latin for "Language of Plants." The original concept for the bar was to base the drink menu on herbal liqueurs and spirits. Preston's previous career had left him with an impressive collection of bottles from around the world. "I was doing a lot of traveling for the defense company," he said, "and everywhere I went I would pick up the local artisan spirits that could only be had there. Invariably that was always some type of an herbal alcohol."

When that employment contract with his former company came to an end, Preston had to abide

by a "strict non-compete" clause for the next five years, which really limited what he was able to do professionally.

Preston's father's family originally emigrated from Spain to the Dominican Republic in the 1800s. His great-grandfather, his grandfather, and his father had all been born on the island. He was the first generation New Yorker in the family, a true pioneer. "I found out I have a huge family down in the Dominican Republic, amazing people. It's a gorgeous island, and I just fell in love with everything down there."

Preston made an amazing discovery while he was visiting. "I was interested in visiting a cacao plantation and one of my cousins said, 'Oh, we have a cacao plantation. I haven't been there in 20 years, but why don't you go take a look?'

"I absolutely fell in love with the place. The trees are magical; they look like something out of a children's book. They're like little apple trees where the fruits grow anywhere on it—right out the side of the trunk. And they're all pastel colors."

At that time, the family was using the fallow farm only for ecotourism, an offshoot of their travel and tourism agency, but Preston had bigger ideas: "I was the crazy cousin who talked them into making a go of it as a production farm. Basically I bought the farm, and then I set out to vertically integrate it." He built a factory next to his bar and began producing Cacao Prieto chocolate and, soon after, making spirits as well. He traded his former career as a rocket scientist with his head in the stars for one firmly rooted on Earth.

In addition to the family's cacao farm (Coralina Farms) in the Dominican Republic, Preston now owns an abandoned limestone mine in Rosendale, New York, whose water he uses to blend his whiskeys, and he also has a partnership in a farm in Hurley, New York, where he grows the grains he uses for his newest whiskeys.

Looking around at this massive, far-flung operation in Red Hook—part chocolate factory, part distillery, part bar, part farm, part laboratory—it is difficult to comprehend how it all came together. "It was the universe conspiring," Preston said with a smile. "We're kind of a crazy assortment of actual rocket scientists and foodies, and we cut no corners. We're super-passionate about what we do and we're fortunate enough to be in a financial position where we can afford to do everything perfectly."

"These bourbons take you on a journey. Like a roller-coaster ride through Eden."

—VINCE OLESON, DISTILLER

THE PROCESS

Preston noticed that in the process of chocolate making, cacao beans needed to be fermented, and what's fermented can be distilled. Preston explained, "Most people don't realize, but cacao—the seed that you make chocolate from—is fermented. If you eat cacao raw, it doesn't really taste like chocolate. The fermentation process is what changes it into the flavors that we think of as chocolate."

Preston started experimenting, producing what could be called chocolate brandy. From there, he began producing chocolate rum and even a dark chocolate liqueur. In 2012, Don Rafael Cacao Rum was the winner of a Double Gold Medal at the San Francisco Spirits Competition, and Don Esteban chocolate liqueur won 90 Points at the Ultimate Spirits Challenge 2012.

The success spurred Preston to look into what he considers "the country's signature spirit." He said, "We're American. We need to try our hand at bourbon." Cacao Prieto set out to make bourbon from non-GMO heritage corns grown in New York State. But that takes time. Meanwhile, he has learned the bourbon business by traveling to Kentucky and handpicking premium barrels of aged bourbon, which he then brought back to Brooklyn and cut to proof with extremely mineral-rich water from an abandoned limestone mine in Rosendale, New York, called Widow Jane. Thus was born Widow Jane Whiskey.

Preston told us, "Going out to Kentucky, you realize that every single distillery treats barrels like gold bullion. The distilleries buy, sell, and trade. The attitude isn't, 'It didn't come from my still, and therefore I can't touch it.' It's raw material. You can pick and choose the best and blend it according to your own taste." The Widow Jane Straight Bourbon, blended in Brooklyn from premium aged stocks, won a Double Gold Medal at the San Francisco Spirits Competition in 2013 and was named by *GQ Magazine* as the top of the best new whiskeys of that year.

Back in New York, Preston was experimenting with his new corn crop. He settled on two heritage varieties called Bloody Butcher and Wapsie Valley, which he grew in great quantities at his farm in Hurley.

Bloody Butcher, one of America's oldest heirloom corns, is a striking deep red in color and is often described as having a "true" corn flavor. The Wapsie Valley corn is a very high protein variety that originated in Iowa and is valued for its use as a grain, so much so that it's finding its way into gourmet products like pancake mix and polenta. Once the corns are harvested, they are brought to Red Hook, Brooklyn, where they are milled, fermented, and pot-distilled at the Cacao Prieto factory. Next, they are aged in oak barrels across the street in their warehouse for one year before being bottled. The results are a new expression of bourbon as we know it, created from heritage American corns grown in New York State, then milled and distilled in Brooklyn.

The two heirloom corns are treated in much the same way, but their tastes are quite different. The Wapsie Valley Bourbon has the inviting flavors of light almond, cacao, and honey notes on the nose. The body is zesty with hints of sour orange peels and sweet praline, and it finishes remarkably smooth for a young whiskey, with lingering hints of mace and honey.

Bloody Butcher Bourbon smells of sweet oak and graham cracker. Though the taste is quite dry at first, it quickly reveals vanilla and allspice notes. The finish is sharp and chewy, full of tobacco and molasses flavors.

WHISKEY FACILE

2 oz Widow Jane Straight Bourbon Whiskey
1 1/4 oz Cardamaro
2 dashes bitters

Build in a shaker, give a quick stir, and strain into a cocktail class as in the photo, or just build over ice and enjoy.

Fred Buscaglione, the inimitable capo of Italian jazz in the 1950s, became enamored with American culture as a child growing up in Italy in the 1940s—so much so that for his stage persona, he adopted a character which was said to be a combination of Mickey Spillane mobsters and Clark Gable. On stage and in film, he aped his favorite subjects, the hardened American gangsters of Prohibition, brutal in their pursuit of infamy but hopelessly at the mercy of women and whiskey. His stage character "Freddy" said he preferred an "easy whiskey" to the ubiquitous mineral water preferred by so many of his countrymen, and added that he drank nitroglycerin for breakfast!

Whiskey Facile is not only an ode to Buscaglione, but also a nod to the preferred drink of seasoned bartenders. At home and off-duty, a bartender will forgo the shaker and ceremony of cocktail preparation, to simply "build" a drink out of a good whiskey and a soft accent like a splash of vermouth or amaro, or a twist of a lemon peel, or a dash of bitters. The accent will depend on mood, or perhaps how to properly follow a meal, or, more simply, on what is available. It's whiskey-easy, whiskey facile.

Danny Kent, bar manager at Locanda Vini e Olii in Clinton Hill, developed this cocktail using a Brooklyn whiskey, an amaro from Buscaglione's native Piedmont region of Italy, and the technique of an off-duty bartender. He suggests: "I think in Freddy's jazzy vein, one should feel free to take this drink and improvise loosely—but whatever you do, never forget to take it easy."

CONQUISTADOR

1 oz Don Esteban chocolate liqueur
1 oz Madeira wine
1/4 oz Pierre Ferrand Dry Curaçao
edible gold flake

*Place the first three ingredients
in a cocktail shaker mostly
filled with ice. Shake and strain
over fresh ice in a rocks glass.
Garnish with edible gold flake.*

This cocktail originated at Botanica, the cocktail bar connected to Cacao Prieto Distillery, as a drink to highlight Don Esteban chocolate liqueur. Don Esteban is pure pleasure for the dark chocolate lovers of the world. Organic cacao beans grown on the distillery's own Caribbean plantation are toasted, then macerated in cane spirit, which produces this deep, dark delight. It goes great alone or pairs well in a number of cocktails.

There is another cocktail called Conquistador that features Irish cream. But last we checked, the Irish were not raiding the New World for its riches, so this version just makes more sense. It celebrates New World discoveries like chocolate and rum and is garnished with gold flakes plundered from El Dorado. Tying these flavors together is an undercurrent of Portuguese Madeira wine. The one-time conqueror is now just part of the mix.

DALE DEGROFF'S BRANDY COCO

1/2 oz Don Esteban chocolate liqueur
1 1/2 oz VSOP Cognac
1/2 oz Cointreau Noir
dash Dale DeGroff's Pimento Aromatic Bitters
1 orange quarter
orange twist

(Optional: frost the rim of the glass with ground cacao nibs, using a light rub from the orange around the rim to get them to stick.) Place the orange quarter in the bottom of a rocks glass and dash with the bitters. Add the Don Esteban and muddle gently to release the juice and oils from the orange. Remove the rind and add ice. Build the rest of the ingredients on the flavor mixture and stir. Garnish with an orange twist.

This drink grew out of a brainstorming session with the incomparable Dale DeGroff. The initial idea was to build a cocktail in tribute to Dale's old friend, the late, great Brooklyn-born singer-songwriter Harry Nilsson. But as Harry was more of a straight spirit, wine, and Champagne drinker (despite being famous for singing a song about de lime in de coconut), Dale wisely opted to go in a different direction, coming up with an elegant way to feature the Don Esteban chocolate liqueur alongside his own eponymous bitters. Dale's Pimento bitters have a classic aromatic bitters profile with a long, drying finish that will elevate many a cocktail to new heights.

> # "Sugar in drinks is like fat in food; it's like butter. It's really important to finish the drink."
>
> **—DEL PEDRO**

PINEAPPLE RYE CRUSTA

1 3/4 oz Widow Jane rye*
1/2 oz lemon juice
1/4 oz simple syrup (*see page 11*)
1/4 oz Luxardo maraschino liqueur
1/4 oz Santa Teresa Rhum Orange
5–6 fresh, medium-sized pineapple chunks
dash Fee Brothers West Indian Orange Bitters

Muddle pineapple in a large metal shaker with bitters, lemon juice, and simple syrup. Add the rye, Santa Teresa Rhum Orange, maraschino liqueur, and ice. Shake well and strain into a 6 oz cocktail glass. Garnish with a fresh orchid.

*The official name is Widow Jane Whiskey Distilled from a Rye Mash, but who wants to take the time to write that?

With this cocktail, Del Pedro of Tooker Alley took his inspiration from the classic Brandy Crusta, an old-fashioned New Orleans cocktail and the first recorded with an over-the-top garnish. He replaced the brandy with a local rye and added muddled pineapple to the mix. For the fancy garnish, the lemon peel has given way to an orchid. The drink was originally named the Crusta for its sugared rim. Del chose to put the sugar right in the drink because he felt it's more important in there to balance the drink. He said, "Sugar in drinks is like fat in food; it's like butter. It's really important to finish the drink."

The Widow Jane rye whiskey is a perfect match for the rest of the ingredients in this cocktail. It's young and grain-forward, and its flavors—which suggest dried apricots and maple sugar—pair wonderfully with the rum.

SWEET JANE

1 1/2 oz Widow Jane Straight Bourbon Whiskey
1/4 oz Amaro Amerigo (*see recipe below*)
dash Fee Brothers Whiskey Barrel-Aged Bitters
good quality or homemade cocktail cherry
(*see page 159*)
orange twist

Place all ingredients in a cocktail shaker mostly filled with ice, stir thoroughly, and strain into a cocktail glass or coupe. Garnish with a cocktail cherry and/or orange twist.

Chris was writing a review of the Window Jane Bourbon for a New Jersey magazine when he created a cocktail in honor of his favorite Velvet Underground song, "Sweet Jane." Obviously, it was a play on the whiskey's name, but it was also a way to describe how the ingredients brought the sweetness out of the whiskey.

Within a week of handing in the review, lead singer and songwriter of the Velvet Underground, Lou Reed, passed away. Reed grew up on Long Island but was born at Beth El Hospital in Brooklyn. He was the unofficial poet laureate of New York City, and this drink—a classic with a DIY twist—is included in this book as our tribute to him. As the man once said, "Between thought and expression lies a lifetime…"

AMARO AMERIGO

16 oz fresh cranberries
2 cups 80 proof vodka
1 cup water
1 cup granulated sugar
3 cinnamon sticks
3 allspice berries
2 whole star anise
1 tsp lemon peel
1/2 tsp grated nutmeg
1/2 tsp fresh ginger

Combine all the ingredients in a jar with a tight lid and set aside unrefrigerated for a week. After a week, open the jar and try the mix. If the flavors aren't strong enough for you yet, reseal and try again in a few days. When you like the taste, strain the mixture, then bottle. Leave alone for a couple days to come together.

Every Thanksgiving, Chris is left with a bag or two extra of cranberries. This recipe came about as a way to boozify that bounty. It's named for Amerigo Vespucci, the cartographer who first surmised the existence of the "New World," and who lends his name to two continents. The recipe makes a bitter, spicy liqueur with an edge of sweetness, or what the Italians call an amaro (Italian for "bitter"). Amaro Amerigo might be enjoyed on its own with ice, or after a meal neat, or mixed with a cocktail recipe in place of things like Campari, Aperol, or even sweet vermouth as in the cocktail on this page. Feel free to adjust the spices and the sugar to your taste

BROOKLYN EGG CREAM

2 oz Don Esteban chocolate liqueur
3 oz local whole milk
2 dashes Coffee-Pecan Bitters (*see page 107*)
seltzer

*Build by placing the first three
ingredients in a collins glass (or ice
cream soda glass if you have one).
Stir, then aggressively add 6 oz seltzer
or to fill and create fun bubbles.*

When Chris was designing a cocktail menu that featured only local booze, this was the first drink he came up with. In fact, the cocktail menu was created to have a home for this drink. Chris has always been a soda fountain junkie, and he thought: What better expression of Brooklyn could there be than an old-fashioned egg cream?

Farmacy, a restored 1920s pharmacy in Carroll Gardens, is our favorite place to get an actual Brooklyn egg cream. Sadly, the place doesn't have a liquor license, so you won't be able to get our adult version there. But the good folks there were still nice enough to let us take this picture in their window.

DER MEXICANER

1 1/2 oz blanco tequila
1 oz Lillet Blanc
1/2 oz Don Esteban chocolate liqueur

*Place all ingredients in a cocktail shaker
mostly filled with ice. Shake, and strain into a
chilled cocktail glass. Sprinkle a pinch
of powdered cayenne pepper on top.*

Author David Wondrich is at the top of the field when it comes to the history of cocktails. His books *Imbibe* and *Punch* are must-owns for any cocktail enthusiast.

He created this recipe—a counterintuitive use of the Don Esteban—for a restaurant called Berlyn, across the street from the Brooklyn Academy of Music (BAM). The result is a stylish tequila cocktail for any time of day.

BEHIND THE WALL

2 oz Widow Jane Straight Bourbon Whiskey

3/4 oz lemon juice

3/4 oz house Winter Syrup (*see below*)

4 dashes Fee Brothers Black Walnut Bitters

a pinch of shaved nutmeg

Combine the whiskey, bitters, lemon juice, and Winter Syrup in a shaker mostly filled with ice and shake. Strain into a rocks glass over fresh ice. Garnish with freshly shaved nutmeg.

This cocktail, originally designed for Lavender Lake's New Year's Eve party, was inspired by the budding bitters collection behind the bar. Bar manager Thomas Callihan said, "We were looking for something to do with Fee Brothers' Black Walnut bitters, and we've been playing around with a winter syrup using allspice, cinnamon, cloves, and star anise. "

The Widow Jane taste profile is that of a classic, aged bourbon. Because it's cut with heavily mineralized water from the Widow Jane limestone mine, it's softer than many bourbons when drunk neat. But here the winter syrup and Black Walnut bitters draw out pumpkin spices and rich toffee notes, elevating the base spirit.

WINTER SYRUP

4 cups water

2 cups sugar (*demerara preferable for richer flavor*)

2 tbsp allspice berries

2 tbsp whole cloves

4 crushed cinnamon sticks

5 whole star anise

Bring the water to a boil in a saucepan. Add in the rest of the ingredients and stir until the sugar is dissolved. Remove from heat and let the ingredients steep for 30 minutes. Strain and bottle. The syrup should last about a month. Add a shot of vodka to increase the shelf life.

We know vermouth isn't a distilled spirit; it's an aromatized wine fortified with spirits, flavored by an infusion of botanicals. If you are anything like us, you have fond childhood memories of foraging through your Grandpa John's liquor cabinet and coming across a dusty, half-empty bottle of Noilly Pratt, or Martini & Rossi, wondering aloud, "What the hell is this stuff? And why does no one ever seem to drink it?!" But after hearing about Uncouth Vermouth, we really wanted to stretch the definition of "spirits" so we could include the company in this book. ▶

Uncouth Vermouth is the brainchild of New Jersey native Bianca Miraglia, who initially crafted her local, seasonal vermouths from a rented corner of the Red Hook Winery (these days she has her own space along Van Brunt Avenue a few blocks away). The name of her company is deliciously clever, and not just because "uncouth" rhymes with "vermouth." The real brilliance of the name stems from archaic definition of "uncouth," when the meaning of the term was closer to "unfamiliar" or "rare," as opposed to the contemporary definition as "rude" or "boorish." The archaic is not only an apt description of vermouth in this era, but it is also the perfect term to capture the wise and wandering spirit of Miraglia herself. When she isn't making vermouth, this artist is creating amazing objects out of wine corks, or foraging in the wilds of Oyster Bay, Long Island for edible ingredients. The silhouette of a Victorian-era woman picking her nose—a prominent feature of Uncouth Vermouth's labels—is a nod to the modern understanding of "uncouth," yet also reflects the playful irreverence of Miraglia, who is as feisty as she is smart.

Miraglia's wine education began with her Italian father, J.P. Miraglia, and continued when her brother Christopher got a job in a wine shop when she was in high school. At 20, she moved to New Hampshire and got a job at a restaurant, The Inn at Thorn Hill. From there, she chose to continue her education by embarking on a Pinot Noir pilgrimage. In 2006, she packed up her car and drove it clear across the United States to Oregon, sight unseen, and with no contacts in the area. She would stay there for three years, returning to the East in 2009, eventually settling in Park Slope in Brooklyn. She said, "I've worked in the wine industry for my entire adult life.

In Oregon, I ended up working with a couple dozen wineries and did everything from assistant winemaking to bartending to managing a distributor to serving in restaurants; I basically did every possible thing I could get my hands on just to learn about wine."

She saw some ugly sides of the business along the way—including an early employer who instructed her to dump a 26-pound bag of Domino sugar into a 1,000-liter tank and then stir it. Miraglia was horrified and remembered, "I'm standing there and thinking, 'This is just terribly wrong.'" Given that experience, it's no surprise that to this day Miraglia refuses to use added sweeteners in her vermouths.

It was Jay McDonald of EIEIO Wines (get it?) who imparted to her a particularly useful piece of viniculture philosophy: "He's the first person who told me that people who refer to themselves as 'wine-makers' are just self-absorbed idiots or chemists. Nobody actually makes wine; wine makes itself."

How did Miraglia make the leap to becoming a commercial vermouth maker? "I decided when I was still living in Oregon that I didn't want to be a winemaker. I'm a control freak. I am not the person who can just sit back and watch it rain the second week in August and hope for the best."

So she broadened her horizons and branched out, hosting underground "pop-up cocktail parties." She explained how one of these clandestine events became the catalyst for her current venture. "We'd get friends of ours who have distilleries to sponsor the booze part of it, and then we'd make vermouths and syrups to tailor the cocktails. And I got really into the

RED HOOK WINERY

Miraglia admires Red Hook Winery's low-key, almost humble approach to the craft of making wine. "The people there are doing some really cool stuff without really squawking about it that much, which I find to be pretty cool. For me, I own the fact that I make a local product that doesn't have any bullshit in it. When most people think about wine, they think, 'Oh, it's wine, it must be natural.' And I very aggressively want to pop that bubble for the general public."

vermouth part of it." She enjoyed the process, saw the need in the marketplace, and set about making vermouth that was different and better than what was available on a large scale commercially. As she puts it, "That's how Uncouth became Uncouth."

Miraglia's mission is to resurrect vermouth from the dusty corners of the liquor cabinet and put it back where it belongs: in people's glasses. "I never wanted to make herbal syrup. I want to remind people that this is wine, and you have to have good wine to make good vermouth—you have to. And that's only the very beginning." Hers is an ongoing quest for perfection, one small batch of seasonal vermouth at a time, all made from local ingredients she has chosen. Each batch is an adventure, unique in its own right. "It's always here and gone; there's nothing I can do about it. I do

everything myself by hand. I'll make a vermouth once, maybe twice a season, and then it's gone and I move on to something else. It's never going to be the same every year. The wine base, certainly, even batch to batch, will always change, just because it's that small. The biggest batch I make is 35 gallons, and that translates to about 24 cases. The most I'll do in a production is two batches. So it's 50 cases maximum, basically, for any production."

All of the wines for Uncouth Vermouth come from the Red Hook Winery's grapes. They are grown Upstate or on Long Island and then turned into wine down on the pier, "from grapes to glass" as she puts it. To fortify her vermouth, Miraglia uses grape brandy made by Finger Lakes Distilling of Burdett, New York, who produces it from New York State wines using New York State grapes.

THE DRY MARTINI

The widespread misunderstanding of the nature of vermouth has been a contributing factor in the "drying" of the Martini. Vermouth once played a vibrant role in the Martini, but the classic cocktail gradually devolved to the point that most bars were serving what basically amounts to a cold glass of gin. Part of that misunderstanding stems from the fact that most people, to this day, tend to think of vermouth as solely a mixing element instead of as a proud beverage in its own right. As the co-star of gin in the context of the classic Martini, vermouth became conflated with the world of spirits, when in reality it belonged elsewhere. Bianca explained, "Vermouth is not liquor. You're talking about something that's under 20 percent alcohol. Bottled properly, it should age like wine, and if you keep it in the fridge after it's been opened, it should live for about a month."

TORINO STYLE "COCKTAIL"

3 oz Uncouth Vermouth (*any variety*)
dash aromatic or citrus bitters (*optional*)
lemon twist

*Fill a glass of your choice with ice
and vermouth, and maybe the bitters.
Express your twist, rub it around the
edge of the glass, and drop it in.*

This non-cocktail recipe is here for a reason. In Torino, Italy, the modern home of vermouth, this is how the regional specialty is enjoyed, as a central part of the nightly tradition known as aperitvo, or, more informally, 'appy 'our. (They're not big on their Hs over there).

Italians do happy hour right. It isn't about pounding beers, sugared cocktails, or greasy plates of fried food. It's a chance to drink a little vermouth and maybe snack on some tramezzini, canapes, pizzette, or perhaps a little smoked fish or pate.

This is one of Bianca Miraglia's favorite ways to drink her stuff and we highly recommend you give it a try.

YONKERS TO BROOKLYN

2 oz Uncouth Vermouth Dry-Hopped
2 oz Nahmias et Fils Mahia
2 dashes Basement Bitters Bitter Frost
(*or other aromatic bitters*)
1/4 preserved lemon slice (*recipe follows*)
float of Galliano

Place the first three ingredients in a mixing glass mostly filled with ice and stir. Strain into a chilled rocks glass, garnish with the lemon, and float a tiny bit of Galliano on top.

When we went to visit Miraglia to take pictures for this book, we noticed a few jars of great-looking preserved Meyer lemons in her refrigerator. We also noticed her impressive collection of cool craft spirits in her home bar. Bianca started playing around and devised this cocktail—sort of a play on the Sidecar—on the fly, for the book. We think it holds up well.

The sweetness of the mahia, a locally produced version of Moroccan fig brandy, matches well with the dry, tannic finish of the hopped vermouth. The Bitter Frost bitters from Tuthilltown Spirits tie the whole drink together, providing a spicy background texture and highlighting the figs in the mahia with heavy flavors of sarsaparilla.

PRESERVED LEMONS

8 organic Meyer lemons
8 oz rye whiskey
1 tbsp maple syrup
1 cinnamon stick
6 allspice berries

Juice and zest half of the lemons. Quarter the remaining lemons and pack them in a jar. Add the lemon zest and juice and remaining ingredients. Close the jar very tightly. Preserve for at least two weeks before use. This amount should fill a 24 oz Mason jar.

WHITE MANHATTAN

1 1/2 oz Kings County Moonshine
1/2 oz Uncouth Vermouth Serrano
Chile Lavender
1 good quality or homemade
cocktail cherry (*see page 159*)

*Place all the ingredients in a cocktail
shaker mostly filled with ice. Stir
until very cold, strain into a cocktail
glass, and garnish with the cherry.*

In a regular Manhattan, the whiskey provides the backbone and the vermouth adds the sweetness. In this white Manhattan, this dynamic is reversed, with the natural sweetness (as opposed to sugar sweetness) coming from the Kings County moonshine and Bianca's Uncouth Vermouth taking center stage in terms of the overall flavor profile. We tried this recipe in a few different iterations and this is the one we liked best.

Nicole Austin from Kings County Distilling is also a fan. "Both elements shine here and they really complement each other. It's natural to pair the spice of the vermouth with the sweetness of the corn." Because the moonshine is by definition un-aged, there is no spice drawn from the barrel in its flavor profile. The serrano chili brings spice to this drink, and the wine in the vermouth brings a tannic element usually pulled from the oak whiskey barrels. It's no wonder the flavors work so well together.

Feel free to play around with this recipe with other Uncouth Vermouths as well.

MARIETTA

(Clinton Hill)

Marietta takes its food and drink seriously, and its owners are as creative as any bar or restaurant in Brooklyn in terms of how they combine and present local spirits. One clear area where they stand out is their ice. Using a combination of unusual techniques and custom cocktail ice molds from Cocktail Kingdom (cocktailkingdom.com), they create a variety of interesting ice preparations, including the one for this drink, which encases cranberries in a collins-shaped cube.

We asked Alicia Blegen where all these ideas come from. "I feel inspired by all the things happening in the kitchen here, and I want to find ways to do things a little better, a little differently. One of the things that we wanted was more cocktail-style ice that would feel a little bit special. This is a neighborhood spot, and people come here for the neighborhood vibe. That's great, but I also wanted to make sure that other people knew that they could come here and have a good time and maybe watch a football game, but also, on occasion, they could have an eating and drinking experience that is, for lack of a better word, transcendent."

CRANBERRY BERET

1 oz Uncouth Vermouth Pear Ginger
1 oz Greenhook Ginsmiths American Dry Gin
1 oz cranberry shrub*
cranberry cubes

Place the first three ingredients in a cocktail shaker mostly filled with ice and shake vigorously. Double strain into a large rocks glass over cranberry cubes.

CRANBERRY CUBES

Use the Cocktail Kingdom Collins Tray. When making the ice, drop a few cranberries into the mold at random. When serving the drink, use a muddler to crack the ice in half. Put both halves in the glass. Blegen described it as "looking like a cranberry bog in winter."

*For a definition of shrubs, see page 101.

There is a lot of ground to cover in this recipe since we have two DIY elements: the cranberry shrub and the unique ice treatment. We loved that Marietta's Alicia Blegen created a drink around two local products. And as if that weren't enough, she named it after a Prince tune...sort of.

CRANBERRY SHRUB

12 oz cranberries

1 cup sugar	1 tsp grated orange zest
1 cup apple cider vinegar	6 cloves
2 cups water	2 whole star anise
1 oz fresh lemon juice	2 cinnamon sticks

Combine all ingredients in a saucepan. Stir frequently over a medium-high flame until the cranberries have been macerated. Strain out cranberry skins and spices using a colander and cheesecloth; be sure to press all the juice back into the mixture using the back of a spoon. Blegen says: "If when cooled the texture is too viscous and thick, dilute the shrub with equal parts cranberry juice and cider vinegar, tablespoons at a time. The desired consistency should be that of a syrup."

EL MEZCÁLIDO

1 oz Uncouth Vermouth Serrano Chile Lavender

1 oz Mezcal Vago

1 oz Aperol

grilled orange slice

Place the first three ingredients in a cocktail shaker mostly filled with ice. Stir until very cold and strain into a cocktail glass. Garnish with a grilled orange slice.

GRILLED ORANGE SLICE

This clever garnish comes from our pal Will Noland of Sidecar. To make the grilled orange slices, simply cut an orange into very thin slices using a knife or a mandolin. Heat a grill or grill pan until very hot and place the slices on it for three minutes or so a side, until the oils release and the outsides darken in color. If you'd like, slices can be preserved in a lidded container with a little vermouth.

The Negroni is one of those templates that every home bartender should feel free to experiment with—we know we have. Bianca's vermouths are a fun variable to play around with in this form, and this creative cocktail, blending flavors of heat, smoke, and flowers, really hits the mark.

UNCOUTH APPLE

1 1/2 oz Uncouth Vermouth Pear Ginger
1/2 oz Industry Standard Vodka
3 oz hot apple cider

*Put the spirits in the bottom of a
coffee cup and fill with the hot cider.
No garnish is needed, but a cinnamon
stick wouldn't be bad.*

We sampled this winter warmer in the cozy confines of Building on Bond during a particularly chilly afternoon. Bar manager Dan Silwinski is a big fan of Uncouth Vermouth, and he spoke to us about why this drink works with Pear Ginger vermouth even though the mixer is hot apple cider. "You have these orchard fruit flavors in this vermouth, along with ginger spice. Bianca uses a high-quality base wine with great acidity. Between the acidity and the bitterness in the vermouth, the high sugar of the cider is balanced out, thus a seamless, balanced cocktail."

Daric Schlesselman is what could be called a moonlighting moonshiner. He maintains his day job as an editor on *The Daily Show with Jon Stewart*, but pursues his passion in his spare time. ▶

One day in the middle of 2011, Schlesselman, along with his wife and business partner Sarah Ludington, rented a 6,000-square-foot Red Hook warehouse. The couple spent the next six months renovating the former paint factory. While they were waiting for their license, Schlesselman played around on a five-gallon copper alembic still that remains in the workspace to this day. They finally got their distilling license on January 1, 2012. They named their distillery "Van Brunt Stillhouse" after Cornelius Van Brunt, an early Breukelen farmer considered by many to have been one of the founders of the borough. How did this television editor, a Wisconsin native who enjoys gardening and home brewing, come to find himself distilling whiskey and other spirits on the cutting edge of the Brooklyn distilling scene? Schlesselman's journey began as a quest to fulfill a need in his life. "I love the creativity and fun of building stories in television,

but I wanted to build a business where my creativity could be directed into a more visceral, tactile product."

At first the idea of distilling wasn't in the equation. "To be totally frank, before a few years ago, I always thought of making spirits as something that was either done by a hillbilly in the mountains or a factory. And I don't have anything against hillbillies in mountains—I come from a long line of them—but that's not what I was looking for. And I certainly didn't aspire to own a factory."

That said, Schlesselman's desire to make something for people to enjoy drinking makes sense when you look at his lineage. "Both sides of my family have been farmers forever. That didn't necessarily foster my love of whiskey, although my family enjoys nice whiskey, but it did definitely instill in me a love of agriculture."

Our own feeling is that part of the craft, organic, farm-to-table phenomenon is fueled by a desire to get back in touch with a simpler life and with more traditional ways of doing things, as a sort of reaction to the challenges of the modern world. Schlesselman agreed, "There's something that feels good about making something yourself. Modern distilling equipment was invented in like 900 AD." At this point, Schlesselman gestured over to his own copper pot still, a beautiful, German-made contraption nearly 12 feet tall. "That still design is from the first millennium. It is not significantly different than the still designs of Scotland, or of the pot still designs of Kentucky. So that's a 1,000-year-old technology. This is an ancient art form, and there's something about gardening, about cooking,

"I always thought of making spirits as something that was either done by a hillbilly in the mountains or a factory. And I don't have anything against hillbillies in mountains—I come from a long line of them—but that's not what I was looking for. And I certainly didn't aspire to own a factory."

—DARIC SCHLESSELMAN

about growing your own fruit trees, that brings us out of the modern technology of the world and into this visceral, tactile life that humanity had experienced up until this last century."

Beyond tapping into that greater sense of how our ancestors lived, what are the benefits to making things by hand in this day and age? Schlesselman's answer started with the obvious and then cut to the very heart of the craft movement in America. "For the most part, they taste better, and they give you a sense of well-being from the fact that you made it, or that your friend made it, or your neighbor made it. And when you are consuming something that was manufactured by somebody completely disconnected from you and packaged in a box and wrapped in plastic, you lose that sense of connection to the thing that makes us live."

Neil Postman, the renowned media critic, was fond of pointing out that technology both giveth and taketh away. In Schlesselman's view, that's definitely the case when it comes to food and beverage production. "On one hand, we can produce hundreds of thousands of cases of beer, or loaves of bread in no time at all, and that is remarkable," Schlesselman says. But something has also been lost along the way. "The modernist period has contributed a paucity of flavors, and that's what drove the craft beer movement, that's what's driving the craft pickle movement, the farm-to-table restaurant movement, and that's what's driving the craft distilling movement."

After all, it's undeniable that a diversity of choices makes life more interesting, and that is as true for people as it is for food products and whiskey.

SMALL BATCH DISTILLATION

A large part of a commercial brand's value, if not all of it, comes from the fact that you know exactly what you're getting. On the craft level, that process can be turned on its head. Given the premium prices one pays for most craft spirits, it can't be a complete surprise every time you open a bottle. There has to be some consistency. Schlesselman is more concerned with a consistency of quality than making sure his products are identical batch after batch, and he uses Due North rum as an example: "Our rum definitely tastes different, batch-to-batch, and I don't try and match the flavor exactly, but I do try and make sure that the last batch is as good as the next batch, and vice versa. I try to keep it in the same flavor profile." You would taste the difference, Schlesselman explained, if you tasted different rums of his side by side, and that's fine with him. This is why he puts lot numbers on the bottles. Not that this approach is easy, as he explained. "I think that's going to be my challenge, communicating that approach to the consumers, but I think that's part of the fun as well."

THE AGING PROCESS

Schlesselman uses smaller barrels for aging his whiskeys in an attempt to get more wood into them faster, but he knows this is not a perfect solution. "There is certainly a part of the aging process that smaller barrels don't address. The aging process is both additive and subtractive. In terms of adding wood flavors, I believe that my bourbon is, without a doubt, equal to an older bourbon. But in terms of the subtractive aspects of aging, Mother Nature takes care of that, and I'm not going to compare myself to an older bourbon in that respect."

APIARY

1 1/2 oz Van Brunt Stillhouse Moonshine

3/4 oz honey syrup (*see page 11*)

3/4 oz lemon juice

2 sprigs fresh lavender (*divided*)

Place the moonshine, honey syrup, lemon, and one of the lavender sprigs in a cocktail shaker mostly filled with ice. Shake well and strain into a chilled coupe. Garnish with the other lavender sprig.

"I did not actually come to distilling from a cocktail point of view, but instead from a sipping spirits point of view." Schlesselman isn't a highbrow artisan who can't stand the thought of someone spoiling his precious elixir by mixing it with other flavor elements; he has just never been into cocktails. "I do like to make some cocktails with the moonshine and the rum. I've made a few variations of Bee's Knees, but I don't like gin, so I've done a Bee's Knees with rum, and I've done a Bee's Knees with moonshine. The moonshine Bee's Knees is sort of my standard go-to cocktail for events these days."

We played around a bit with Schlesselman's recipe for a moonshine Bee's Knees and rechristened it The Apiary. The moonshine is rounded out with a warm honey note, some brightness from the lemon and a nice floral element as well.

SLEEPYTIME TRIO

2 oz Chamomile-Infused Moonshine (*a.k.a. Camodog*)

3/4 oz lemon juice

1/4 oz honey syrup (*see page 11*)

1/4 oz cocktail cherry syrup (*either from good commercial cherries or the recipe on page 159*)

Put all ingredients in a cocktail shaker mostly filled with ice. Shake and strain over fresh ice in a rocks glass.

CHAMOMILE-INFUSED MOONSHINE

To make the Camodog, add four bags of chamomile tea to a 750 ml bottle of moonshine, and steep at room temperature for 45 minutes to an hour (note that booze will extract a lot of flavor out of the tea bags much more quickly than water). Make sure to squeeze the goodness out of the teabags when you pull them out.

Will Noland of Sidecar is a perfectionist when it comes to the important things in life…like making drinks. "I'll do seven different versions of the same drink and only a total obsessive nerd could tell you the difference between them, but that's OK—obsessive nerdiness is the basis of all craftsmanship."

This drink—named after Noland's favorite hardcore band, Sleepytime Trio—takes a little bit of time. But we like how the addition of the chamomile plays with the moonshine. It doesn't so much mellow it—who'd want to do that? —as it just gives it a floral and oily character.

SOUR CHERRY COCKTAIL

2 oz Van Brunt Stillhouse American Whiskey
1 oz fresh lemon juice
1/2 oz simple syrup (*see page 11*)
1/2 oz homemade sour cherry shrub

Put all the ingredients into a shaker with ice.
Shake, then strain into a rocks filled glass.

SOUR CHERRY SHRUB

1 pint sour cherries
1 pint sugar
7 ounces white vinegar
1 ounce balsamic vinegar

Stir together the fruit and sugar. Let stand for at least a day
in your refrigerator. The mixture will start to get syrupy.
Press the solids through a sieve into a mixing bowl, trying to
extract as much liquid as possible. Scrape any excess sugar
into the bowl as well. At this point, add the balsamic vinegar
and about half of the white vinegar (you don't always need
the full amount of white vinegar). Whisk what you've got in
the mixing bowl until the sugar is combined. Now you have
to taste. Is there enough sweetness? You can always add more
sugar if not. Is there enough tartness? That's where your extra
vinegar comes in. You definitely want the shrub to be tart, but
you're looking for homemade lemonade tart, not salad dressing
tart. Use your best judgment. Once you're happy with the
flavor, bottle it, and you're good to go.

Shrubs were originally a Colonial-era beverage borne out of the desire to preserve fruits past their natural expiration dates. Shrubs are made by combining fresh fruit, vinegar, and syrup. They make a delicious non-alcoholic base for homemade sodas, they are delicious on their own, and they have become a very cool cocktail ingredient as well.

Will Noland of Sidecar makes shrubs regularly. He can often be found in the kitchen on Monday afternoons tasting, blending, and straining. "I'm a country boy. And growing up in the country, I can remember my mom canning tomatoes. She'd wait for these beautiful tomatoes to get to the exact moment when they were at their most perfect and that's when she'd can them—preserving that flavor for the whole winter. That's the idea behind making shrubs."

This is the whiskey cocktail Noland makes with the shrub.

CANNONBALL

GUEST COCKTAIL BY ST. JOHN FRIZELL, OWNER OF FORT DEFIANCE

I came up with the name for this drink first. I always wanted to name a drink the Cannonball, because a cannonball, possibly from Red Hook's Fort Defiance, was found near the Fort's original site back in 2006. And there's a funny story about the Battle of Brooklyn and an ill-fated pet iguana named Geoffrey. But also just because naming drinks after weapons is cool.

I knew flavor-wise it had to harken back to Revolutionary War days, when gentlemen drank lots and lots of punch. I knew I wanted it to contain rum (very popular all over the northeast in those days), green tea (a common punch ingredient, also popular at the time), and pineapple (a colonial ingredient—if a rare and coveted one—and delicious with rum). The rest was trial and error—finding the right proportions and the right ingredients to round out the first three. Turns out a dash of Fernet Branca worked wonders, adding a grounding bitter note and a touch of mint. I call it a "tropicolonial" drink, combining elements of colonial and tiki-drink theory. Using Schlesselman's Due North was a given, and it turns out that infusing his rum with pineapple brings out a yummy toasted coconut flavor in the liquid that I had missed previously.

2 oz Pineapple Green Tea-Infused Due North Rum
1/2 oz Lemon Hart 151 Demerara Rum
1/2 oz Kalani Coconut Liqueur
1/2 oz fresh lime juice
1/4 oz Fernet Branca

Combine all ingredients in a cocktail shaker mostly filled with ice. Shake and pour over fresh ice into a rocks glass. Garnish with a lime and a cocktail umbrella.

PINEAPPLE GREEN TEA-INFUSED RUM

2 cups Due North Rum
1 medium pineapple
1 green tea bag (we use In Pursuit of Tea's Hojicha)

Peel and core the pineapple and cut into spears. Place in a large Mason jar with the rum for approximately two days. After the two days, add one bag of green tea and leave for another day. Strain the solids out and discard.

APPLEWOOD

(Park Slope)

As we were chatting with Brendan Casey from applewood about the joys of the Van Brunt Stillhouse and the different ways of barrel-aging cocktails, a whole pig, ready for butchering, made its way from the walk-in freezer to the kitchen. Clearly this is a restaurant that takes its farm-to-table approach seriously. We asked Casey if the restaurant itself was an influence on his drink-mixing. "It absolutely is," he answered. "Seeing everything that goes on in the kitchen, how everything evolves with the seasons and how different flavors take center stage at different times of the year, is something I want to reflect in our cocktail program."

AMERICAN JOE

1 1/2 oz Van Brunt Stillhouse
American Whiskey
1/2 oz honey syrup (*see page 11*)
1/2 oz Ramazzotti
1/2 lemon juice
dash Coffee-Pecan Bitters

*Put all ingredients in a cocktail
shaker mostly filled with ice.
Shake and strain over fresh ice in
a rocks glass. Top with club soda
and garnish with a lemon twist.*

Brendan Casey, who ran the beverage program at applewood for many years, designed a cocktail for this book using Van Brunt Stillhouse's American Whiskey and a homemade bitters recipe he got from Brooklyn resident Brad Thomas Parsons' must-have cocktail/DIY book, *Bitters*. Parsons, a major influence on this book himself, was kind enough to let us reprint the recipe.

As for the cocktail, Casey described its unique combinations in a way that made us thirsty. "There's a real harmony between the elements here. The Coffee-Pecan Bitters complement the nutty quality in the young whiskey, and the Ramazzotti has spicy notes that also work well with the whiskey. The flavors all come together in a way where they really showcase the whiskey and bring out the finer notes in it."

COFFEE-PECAN BITTERS

BY BRAD THOMAS PARSONS, AUTHOR OF *BITTERS*

2 cups high-proof bourbon, or more as needed

1 cup water

1/2 cup lightly toasted pecans

1/2 cup whole coffee beans, lightly cracked using a mortar and pestle or the bottom of a roasting pan

2 tbsp sorghum syrup or molasses

1 tsp cacao nibs

1/2 tsp cassia chips

1/2 tsp wild cherry bark

1/4 tsp dried orange peel (to make, bake orange zest at 200°F for 30 minutes on a baking sheet)

1/4 tsp black peppercorns

Place all of the ingredients except for the bourbon, water, and sorghum syrup in a quart-sized Mason jar or other large glass container with a lid. Pour in the two cups of bourbon, adding more if necessary so that all the ingredients are covered. Seal the jar and store at room temperature out of direct sunlight for two weeks, shaking the jar once a day.

After two weeks, strain the liquid through a cheesecloth-lined funnel into a quart-sized jar to remove the solids. Repeat until all of the sediment has been filtered out. Squeeze the cheesecloth over the jar to release any excess liquid and transfer the solids to a small saucepan. Cover the jar and set aside.

Cover the solids in the saucepan with the water and bring to a boil over medium-high heat. Cover the saucepan, lower the heat, and simmer for 10 minutes.

Remove the saucepan from the heat and let cool completely. Once cooled, add the contents of the saucepan (both liquid and solids) to another quart-sized Mason jar. Cover the jar and store at room temperature

out of direct sunlight for one week, shaking the jar daily.

After one week, strain the jar with the liquid and solids through a cheesecloth-lined funnel into a quart-sized Mason jar. Repeat until all of the sediment has been filtered out. Discard the solids. Add this liquid to the jar containing the original bourbon solution.

Add the sorghum syrup to the jar to incorporate, then cover and shake to fully dissolve the syrup.

Allow the mixture to stand at room temperature for three days. At the end of the three days, skim off any debris that floats to the surface and pour the mixture through a cheesecloth-lined funnel one last time to remove any solids.

Using a funnel, decant the bitters into smaller jars and label. If there's any sediment left in the bottles, or if the liquid is cloudy, give the bottle a shake before using. The bitters will last indefinitely, but for optimum flavor use within a year.

— *Reprinted from Brad Thomas Parsons'* Bitters *with permission of the author.* —

BARREL-AGED OLD-FASHIONED

750 ml Van Brunt Stillhouse Moonshine
750 ml Van Brunt Stillhouse American Whiskey
750 ml Van Brunt Stillhouse Bourbon
10 dashes Angostura aromatic bitters
10 dashes Peychaud's Bitters
10 dashes Regans' Orange Bitters No. 6
8 ounces simple syrup (*preferably made with Demerara sugar, see page 11*)

Using a bowl large enough to hold all the liquid, combine the above-listed ingredients. You'll want to taste at this stage to see if the amount of sugar and bitters is to your particular taste.

Using a funnel, pour the cocktail from the bowl into the barrel. Feel free to take the intermediate step of moving the cocktail from the large mixing bowl into a pitcher with a spout. Once all the liquid is in the barrel, seal the barrel and put it in a dark, cool place. Then it's time to let the waiting begin.

Says Casey, "You'll want to taste it frequently. Maybe start after a week or two. Usually I've found that six months is a sweet spot where all the flavors come together if you can wait that long. There are a lot of variables and it's fun to try it along the way."

The barrel itself is a big variable. If it's been used to age something else, you'll get much less wood and more of whatever flavors were in there from the previous use. Tasting a lot is the only way to know what you've got—especially with a new barrel. You don't want that barrel flavor to become overwhelming.

Barrel-aging cocktails is all the rage, and with three-liter barrels currently available from a number of sources online in homebrew stores, and even from Van Brunt Stillhouse itself, it's never been easier for the home bartender to try his or her hand at barrel-aging a batch of cocktails.

This recipe, shared with us by Brendan Casey, is a great place to start. If you're interested in the concept here but don't want to deal with an actual barrel, bottle-aging is a viable alternative. Obviously, you won't get the taste fluctuations you will with barrel-aging, but you'll still see flavors meld together and change over time.

We asked Casey about the merits of aging cocktails. "Aging changes the context of what the cocktail is. Time helps integrate and meld the flavors. If you continue to taste over time, you'll notice that the flavors move apart and come back together at certain points."

And how does that process work with the spirits in this combination?

"The idea is to solidify and sweeten the flavors. In barrel-aging an Old-Fashioned, you're going to be adding woody flavors like vanilla that will also add complexity and mellow the whiskey."

This recipe combines three spirits made at Van Brunt Stillhouse with bitters and sugar. The result is a memorable concoction that will teach you a lot about what you're drinking as you taste it along the way.

You can downscale the recipe by quartering the amounts for bottle-aging if you'd like. The amounts listed are for use with a three-liter barrel.

DUE NORTH MANHATTAN

2 oz Due North Rum

1/2 oz plus a splash* of tawny port

2 splashes Ramazzotti

2 splashes sweet vermouth

cocktail cherry (*page 159*) with a little bit of its syrup

Place all ingredients besides the cherry in a mixing glass mostly filled with ice. Stir until very cold and strain into a cocktail glass. Garnish with the cocktail cherry.

Because it is aged in charred new oak barrels, Due North rum has some whiskey-like properties, so it isn't a surprise that it works well in a drink like the Manhattan. Schlesselman makes a version of this one that he calls the Battery Tunnel. Over at The Good Fork, Matty McDermott makes this iteration, where he uses various elements, including tawny port, to round out the drink nicely and make it his own.

*How much is a splash? A very small amount, maybe 1/32 of an ounce. Use your own taste and judgment to guide you in this preparation.

SNOW BLIND

2 oz Due North Rum
1/2 oz lime juice
½ oz simple syrup (*see page 11*)
clove studded lime wheel
star anise

*Place the first three ingredients
in a mug and top with hot
water. Garnish with the
clove-studded lime wheel and
float a star anise pod on the top.*

The winter of 2013–14, when this book was being written, was just brutal in terms of weather. Countless shoots and interviews were juggled and rescheduled because Mother Nature just would not cooperate. There were very few times that we were glad to be writing this book over such a long, lousy winter, but one of those rare times came when we tasted this cocktail from Lavender Lake—it's a local take on a Hot Toddy, but the flavors and aromas are transporting.

Conrad Oliver explained how this drink takes on a life of its own on a cold day. "First one person orders it, then a dozen people will order it. It always does really well if the weather is like this."

Creator Thomas Callihan loves the Due North Rum in this one. "The rum is aged in charred whiskey barrels so it brings those spicy, oaky, vanilla flavors as opposed to what you'd expect from your average rum. It gives a funky, Brooklyn flavor to it."

"Obsessive nerdiness is the basis of all craftsmanship."

—WILL NOLAND

WHISKEY APPLE

2 oz Van Brunt Stillhouse Bourbon
3/4 oz Apple Simple Syrup
1/2 oz lemon juice
dash kümmel
dash of your favorite aromatic bitters

*Shake all ingredients hard in a cocktail
shaker and strain into a rocks glass.
Garnish with a dried apple chip.*

NOTE ON APPLE SIMPLE SYRUP:

*"I tried a million ways of making apple syrup, and in
the end, none of the crazy stuff I was doing tasted as
good as just taking equal parts of a good apple cider
and sugar and mixing them until they combined."*

NOTE ON DRIED APPLE CHIPS:

*Slice an apple thinly with a mandolin or sharp knife.
Place on a drying rack in the middle of a 175° F oven.
Baking time will depend on the width of the slices, but
there should be just a little pliability to them that will
dry out as they cool. Let them cool on the rack but pull
them loose while still pliable or they'll stick to the rack.
As Noland says, "You can cheat it to 200–225 to cut
the time down, but they won't look as pretty."*

Our attraction to Sidecar was pretty straight-forward when it first opened in 2008. It was one of the few places in the vicinity of Park Slope that was open late. Then we tasted the brandade and our appreciation of the place began in earnest. There has been a good cocktail program in place at Sidecar since it opened, but in our view, the arrival of DIY cocktail expert Will Noland has taken the cocktails at Sidecar to a new level. Whether it's his own wacky creations like the Spicy Jesus or his take on classics like the Bramble (see page 148), Will has a great palate and a chef's ingenuity when it comes to cocktails.

This spirit-forward gem is a great way to enjoy any of the Van Brunt Stillhouse whiskeys, particularly the bourbon, which we used for this drink on the day we shot the accompanying photo. The spices in the kümmel (caraway seed, cumin, and fennel) help bring the sweet flavor of the New York State corn in the bourbon front and center.

How did Kentucky native Colin Spoelman find himself in Brooklyn, crafting a libation much more commonly found—not to mention less expensively produced—back in his home state? "I grew up in Kentucky, but I didn't have any intention of starting a distillery. And even after moving to New York, I worked in film; I worked in Estee Lauder in the perfume department for a little while." That odd job would actually prove quite significant in his future as a distiller, because working in the perfume department forced Spoelman to pay attention to his sense of smell, to train himself to recognize subtleties and nuances that others might miss. ▶

His Kentucky heritage ultimately did play a role in his choice of business. "I didn't know anything about moonshine. I didn't know anything about whiskey. For me, I started making whiskey just to have fun, and to participate in this cultural heritage that I still felt very connected to, even though I lived in the city. It was a way to connect to the culture that I was raised in.

"In a global culture, cities have always been the places that you can move to if you want to forge a new identity. But I also think there's a dissociative aspect of the city. When you move to a city as diverse and culturally rich as New York or Brooklyn, there's a little bit of something lost in terms of where you came from. I landed here and was forced to contend with how my culture related to this broader culture. A lot of people, I think, turn to what they came from, and in my

case that meant whiskey. So I would say what's happening in Brooklyn isn't any different from what has ever happened in Brooklyn, which is that people come here to make a new life and apply their skills and creativity to doing something that then transports the culture of where they came from to here."

Spoelman started out by making moonshine illegally, and this hobby grew into Kings County Distilling. He had two "eureka" moments during that growth process. "The moment that launched the company was when we were doing a blind tasting of white whiskeys, with people that I didn't really know, around the holidays in 2008. And people preferred the stuff that I was making above what was commercially available. That was all the confirmation I needed that being in the distillery business was a good idea."

The other moment came later on, when Spoelman sent in samples of his bourbon for chemical testing, along with samples of two popular, national-brand bourbons. The national brands came back at 40-50 parts per billion of methanol (a nasty, hangover-worsening chemical), while Kings County's own craft bourbon registered only 5 ppb. "For me that was a very important moment," Spoelman said. "I already believed that what I was making tasted better. I didn't have any quantitative evidence that it was actually better, but to see that on paper in front of me was the confirmation that I was looking for."

Spoelman, along with his partner in the distillery, David Haskell, have penned a book about whiskey called *A Guide to Urban Moonshining: How To Make and Drink Whiskey*. The book is an excellent read—part history, part memoir, part guidebook, part manifesto.

In the book, Spoelman presents a new way of looking at the world of spirits, one much more focused on what's in your glass and how it got there than on how it's been marketed. "I implore people to go out and taste their whiskeys blind, because you'd be very surprised what you'd like. We want people to think differently about what they consume."

It is clear that Spoelman has a good perspective on the cultural shift towards craft goods and DIY. "Twenty years ago, people viewed huge corporations as being a very safe alternative to the farmer down the road. 'It must be safer to eat the corporation's chicken,' they reasoned, 'because it's made by a huge company that has safety requirements, and ultimately I'm going to be healthier because I'm eating that.' But I think in the last 20 years people have swung around to the inverse of that, and have become very suspicious of large corporations, and the fact that there isn't really transparency, and there are a lot of shortcuts, and you can never really be sure that what they're making is better. Maybe the farmer down the road was doing something right all along."

> ## "Maybe the farmer down the road was doing something right all along."
> —COLIN SPOELMAN

WHAT MAKES GOOD WHISKEY?

Too many people have internalized the myths surrounding whiskey: the types of water that can be used, the kinds of barrels required, how long it needs to be aged, and, of course, those geographic restrictions. Spoelman chuckled as he explained, "I think that's one of the things that people are surprised about—that you can make very tasty whiskey outside of Kentucky, and you can make very good whiskey that doesn't have to age for 12 years. And you can make whiskey without limestone water, and you can make it from grain other than corn."

The answer to the age-old question, "What makes good whiskey?" really comes down to the use of high-quality ingredients, "pot distillation, and narrow cuts from the still as you're distilling, which gives you a cleaner but equally robust spirit," according to Spoelman. For Spoelman, the "front side cuts tend to be very solvent-y and kind of like nail polish remover. They have that very astringent, chemically-laden sort of flavor." Contrast that with the cuts on the back end, which have "more of a pungent, aromatic quality and are fruity, in a somewhat artificial way. There are a lot more esters on the backside, and the esters tend to give it this artificiality." It is the cuts in the middle of the distilling run that are generally considered to be the best: "They're the cleanest and...the most familiar to the nose, and that's where the best whiskey is." Commercial distilleries, Spoelman says, face this dialectic tension between taking the narrower cuts for the best flavor, and financial pressures that can lead them to cut too widely; the wider cuts lead to more offensive, chemical elements finding their way into the final blend.

Aging, of course, plays a role in making whiskey, Spoelman says, and there are a number of factors that influence that phase of whiskey production: the temperature in the barrel storage room, the type of wood used, the "char" of the barrel, the size of the barrel—all of these things have an impact on how the whiskey will taste. Perhaps the greatest myth of all is that the longer the whiskey is aged, the better it will be. Not true, says Spoelman. The real trick is to find the whiskey's "sweet spot." Taken to the extreme, at a certain point you are more drinking the barrel than the whiskey. "Another myth is that if you were to drink the 100-year-old whiskey it would be better than the 10-year-old whiskey, and that would be better, in turn, than a one-year-old whiskey. But it just depends on what the conditions are. In the same way that wine has a balance point, after which it will start to deteriorate, I think whiskey, too, has a sweet spot, and there's a novelty to that. I like the opportunity to drink very old whiskey, just because it's rare, and unusual, and it teaches me about the distilling process. But is it better whiskey? I would certainly argue that an 18-year-old Macallan is better than a 25-year-old Macallan. It goes past the point of balance and goes too heavily into the wood category."

"We want people to think differently about what they consume."

—COLIN SPOELMAN

MOONSHINE

For Spoelman, nothing demystifies making whiskey quite like understanding the world of moonshine. As city dwellers, we grew up with the persistent cultural stereotype of the toothless, flannel-clad backwoodsman hiding his rusty still in the forest, producing an almost unpotable concoction that could render one blind or insane with a single swallow!

Moonshine is whiskey, often referred to as "white whiskey." Spoelman uses the terms interchangeably, for moonshine is essentially whiskey before it goes into a barrel. He pointed out that in a category like tequila, for example, people don't give a second thought to the fact that you can buy both aged and un-aged versions at your local liquor store; the same holds true for rums and brandies. Most people don't even make a distinction. Spoelman says, "They go to the bar and they order a tequila. They don't say, 'Oh my God, I would never drink un-aged tequila, that's horrible.'

"Some people think that white whiskey is a fad, when, in fact, in the long history of whiskey, there's probably been more white whiskey consumed than aged whiskey. And for me, when I got started making whiskey, I was trying to answer the question, 'What makes good whiskey?' And initially I didn't have the resources to age the whiskey, so it was much easier for me to negotiate those variables when I was making white whiskey. I could make a very good, palatable white whiskey, and then sort of knew, by extension, if you put that into the barrel that it would end up being a good spirit. I would argue that to learn more about whiskey, don't bother chasing down these very expensive bottles of Scotch that are really old. Drink more white whiskey, because then you'll actually learn what it is you like from the spirit."

Spoelman used rye as an example: "A lot of people use the tasting note 'spicy' when describing rye, as opposed to Scotch, or wheat whiskey, but if you were to drink the 'white,' un-aged version of the rye, you would realize that this 'spiciness' does not inhere in the grain itself. You would then understand that the 'spicy' flavor note is a result of the rye's contact time with the barrel."

NICOLE AUSTIN

One of the most respected figures on the Brooklyn distilling scene is Kings County Master Blender Nicole Austin. In a few short years, Austin has put herself in the middle of a craft distilling boom in New York State, acting as distiller, lobbyist, blender, and good will ambassador at large.

Austin admitted that she began drinking whiskey in high school because she "thought it made her look like a badass." That early attitude led her eventually to leave a successful career in environmental engineering to pursue the art of distilling.

She originally joined KCD in August 2010 as an unpaid employee, just eager to get her feet wet in the business. Today she's a partner at the distillery. It's her responsibility to taste and blend all of the whiskey that comes out of the distillery. That means that she is deciding what you drink.

Her goal in that endeavor is clear. She says she "wants people to feel a connection" to both their experience and their story. For her, making good, local whiskey is very personal. Austin sees a day when regional whiskeys can stand out from each other due to local grains, water, and even technique. She's hinting at the idea of terroir that is discussed in wine. If it's possible for something to be defined as a "New York" whiskey, Nicole will be part of it.

In addition to influencing the tastes of Brooklyn whiskey drinkers, she is also helping shape the New York State laws that regulate the craft distillers. As president of the New York State Distillers Guild, she is a go-between for lawmakers and distillers and has a great part in shaping the future of the local distilling industry.

How does she like her new, challenging career since leaving environmental engineering just a few years ago? "I feel like I haven't gone to work for a single day since I left my old job."

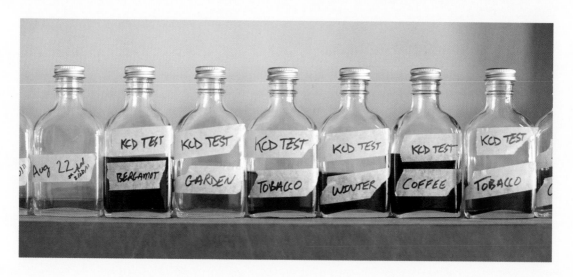

KINGS COUNTY SOUR

1 3/4 oz Kings County Bourbon

1 egg white

3/4 oz lemon juice

1 bar spoon sugar

1 splash of port wine

orange slice

good quality or homemade cocktail cherry

(see page 159)

Shake the first four ingredients hard in a cocktail shaker first without ice, and then with ice. Strain into a rocks glass over fresh ice. Add a little port wine float. Garnish with an orange slice and maraschino cherry like Luxardo or the homemade version on page 132.

This is a drink for people who really want to taste whiskey in their whiskey sour. Kings County Bourbon has a strong platform of sweet, silky corn, which adds toothsome undertones of light brown sugar, caramel, cinnamon, and a touch of orange peel to the cocktail. The finish benefits from the bourbon's 18-month-long barrel-aging, giving it a spicy throat-feel, which is not too overwhelming, but definitively the burn of bourbon.

Henry Public co-owner Matt Dawson says of the drink, "It's named after the county of Brooklyn. It's our variation on the classic New York Sour. Instead of red wine, we float a little port on our whiskey sour. And we thought port made sense since we're just a few blocks from the piers. This is the Port of Brooklyn after all."

THE BUCK STOPS HERE

2 oz Kings County Bourbon

4 oz Pickett's Ginger Beer

(made from Pickett's ginger syrup)

3 dashes Hella Bitter aromatic bitters

1 wedge orange

In a highball glass, muddle fruit and the peel of orange wedge. Top the glass with ice and bourbon, bitters, and ginger beer. Pour back and forth from glass to shaker a couple of times to properly mix. Serve with a straw.

Noorman's Kil is a whiskey and grilled cheese bar in Greenpoint that derives its unusual Dutch handle from the original name for Bushwick Creek. The place has an impressive array of over 400 bottles from around the world, including several from Brooklyn. It also has a small but well-curated cocktail program featuring spirit-forward drinks. Another thing that makes the cocktails unique here is the exclusive reliance on local Hella Bitter bitters (see page 134).

We had a chance to invade Noorman's Kil to hang out with Team Hella Bitter and Noorman's Kil co-owner Harold Simoneau during our research. We messed around with a few cocktails, mostly designed by Hella's Tobin Ludwig, made with local booze. This one was a clear winner—we wouldn't be surprised if Simoneau puts it on the menu soon.

HENRY PUBLIC

(Cobble Hill)

Matt Dawson is a co-owner at Henry Public near Atlantic Avenue, just down the block from Toby Cecchini at Long Island Bar. Henry Public is a stylized spot, but still cozy. It's designed to look like an early 20th-century tavern, not a speakeasy. And as you can see from the pictures, it's got a great look. About the inspiration behind his cocktail program, Matt says, "Many of our cocktails are inspired by classics—our own twists or variations on old-time classics from the days before, during, and after Prohibition. We like drinks that are delicious, balanced, and not too fussy. We like drinks that go with food."

The food item Henry Public is best known for is the turkey leg sandwich. We asked Matt what cocktail goes best with it. He responded, "What cocktail *doesn't* go well with a turkey leg sandwich? I suppose anything brown-spirit-based would be best: a Manhattan, our Brooklyn Ferry, Ward Six, the KC Sour all come to mind."

Get to Henry Public as soon as possible to conduct a taste test of your own.

DARK CHOCOLATE OLD-FASHIONED

1 1/4 oz Flor de Cana 7yr Rum
*(another aged rum can be an able
substitute here)*
3/4 oz Kings County Chocolate
"Flavored" Whiskey
dash simple syrup *(see page 11)*
2 dashes Fee Brothers Peach Bitters

*Place all ingredients in a mostly
ice-filled cocktail shaker, stir,
and serve over fresh rock(s),
with an orange twist.*

Sweet Chick is known for chicken and waffles, but it also has a cool cocktail program with plenty of local offerings. This drink was designed by bartender Anthony Sferra. Like Dale Degroff does with his Devil's Due cocktail (see page 203), Anthony added a little aged rum to round out the young Kings County Chocolate "flavored" Whiskey. Bartender Jeff Sims, pictured here, made the one we photographed.

Kings County Distillery makes its "chocolate" whiskey by infusing its moonshine with the husks of cacao beans sourced from Brooklyn's own Mast Brothers Chocolate. The husks impart bitter chocolate aromas and flavors to the moonshine, which are reminiscent of rich truffles when they meet the sugar and grain in the body of the base spirit. The finish is pure, unsweetened baker's chocolate.

You can try this cocktail in a variety of ways, just as we have it here or with all-chocolate whiskey. It's a bit intense the latter way, but if you're a fan of cacao, you'll love it.

Playing with the bitters in this recipe is a fun experiment as well. That's the great thing about the Old-Fashioned—it's a good template for the home bartender to play around with.

GRAVESEND SMASH

2 oz Kings County Moonshine

3/4 oz St Germain

1/4 medium sized grapefruit

1/2 tsp pink peppercorns, divided

Muddle half the peppercorns and grape-fruit with the St. Germain in the bottom of a cocktail shaker. Add the moonshine and fill the shaker mostly with ice. Shake vigorously and strain into a collins glass filled with fresh ice, small cubes or crushed. Garnish with the rest of the cracked pink peppercorns on top.

All juleps are smashes. But not all smashes are juleps.

Both drinks usually combine sugar and/or fruit and a base spirit (though not always, as you'll see with One Mint Julep, page 132). As far as we can tell, the thing that most makes a julep a julep is that it's served in a julep tin and always has cracked ice.

This smash is a deranged younger brother to the Mint Julep, who dropped out of college to spend time cycling and picking wildflowers in Dijon, France…or something.

The reason we say "younger brother" is that Kings County Moonshine is un-aged corn whiskey—in other words, you could put it in a barrel and make whiskey. This is the essence of spirit making—a pure expression of corn. The nose is light with a soft, earthy sweetness of spring grass. The sweetness in the body is reminiscent of popcorn shoots: grassy and light with a wisp of honey syrup. Because the shine is unfiltered, there's a pleasant sheen of texture that helps to elevate and pronounce the lighter flavors on the palate. The finish is sweet, too, but does not linger and there is no burn.

We love the combination of elements in this cocktail: the gentle spice and pleasing aesthetics of the pink peppercorns; the way the muddled grapefruit mellows the un-aged whiskey; the floral quality of the St. Germain elevating the individual elements.

As for the name, Gravesend was one of the old racetracks in Brooklyn, the birthplace of the Brooklyn Derby, located between what is now McDonald Avenue and Ocean Parkway, from Kings Highway to Avenue U. Gravesend launched in 1886 and was crushed by the calamitous Hart-Agnew bill in 1908 that effectively banned all racetrack betting in New York State. Perish the thought. Pete believes that, whenever possible, drinks should be named after horses, stakes races, current racetracks, or failing that, defunct racetracks.

ONE MINT JULEP

2 1/2 ounces Kings County Bourbon

4 sprigs mint

crushed ice

seltzer

Frost a silver cup in your freezer. Remove the stems from three mint sprigs and place the leaves in the bottom of a silver cup. Add a splash of bourbon and muddle the mint. Add the rest of the bourbon, pack the cup with crushed ice, and then top with a little bit of seltzer. Garnish with the remaining mint sprig.

Kentuckian Henry Watterson famously wrote the following recipe for the Mint Julep. "Pluck the mint gently from its bed, just as the dew of the evening is about to form upon it. Select the choicer sprigs only, but do not rinse them. Prepare the simple syrup and measure out a half-tumbler of whiskey. Pour the whiskey into a well-frosted silver cup, throw the other ingredients away and drink the whiskey."

This is Colin Spoelman's kind of cocktail recipe. But even Spoelman, who has acknowledged a certain "cognitive dissonance" when it comes to making cocktails with his whiskeys, understands the importance of having a Mint Julep on the first Saturday in May. As a native Kentuckian, Colin measures the passage of time by the Kentucky Derby. Who needs New Year's Eve when you have the most exciting two minutes in sports? Pete feels the same way and often refers to the year he met his wife Susan not as 1997 but as "the year Silver Charm won the Derby." Here is a julep recipe from Colin that gives his bourbon its due—front and center in this Kentucky classic.

HELLA BITTER

Eddie Simeon's first job was working as a bar-back in Oakland, California. He was underage at the time and ended up drinking a lot of bitters and soda because it was one of the few decent, non-alcoholic beverage options.

When Eddie moved to Williamsburg, Brooklyn in 2007, he and his roommate concocted their first batch of homemade bitters, guided by an online recipe. It took them two days of scavenger hunting just to find all of the ingredients for a Mason jar-sized yield. The boys had to eat hot sauce for an entire week just to have enough jars to store it in!

The hobby grew. In 2011, Hella Bitter did a Kickstarter campaign to raise funds for a large batch of pre-paid bottles. When the campaign netted them 250 percent of their modest $1,000 goal, they began to think that

maybe this could become a real business. They took their surplus product to local retailers such as Whisk and Meadow and to local bars and restaurants.

They went from making five-gallon batches to 30-gallon batches overnight, and once they made the move to a Queens production space and rented production space from Organic Food Incubator, they had two, 550-gallon tanks at their disposal. Considering that this is a product generally measured out by eyedropper, that's a lot of bitters! Today the company consists of Simeon, his high school buddy Tobin Ludwig, and their colleague Jomaree Pickard. They each bring something useful to the table, with Eddie functioning as the sales manager and strategist, Tobin as head of production, and Jomaree, an alumnus of the Wharton Business School, as resident business wiz.

Hella Bitter currently offers two products: Citrus and Aromatic. It sells the standard five-ounce bottles, and also offers 1.7-ounce bottles, with both flavors packaged together and nicknamed the "Salt and Pepper Pack," in keeping with the concept of bitters as a "bartender's spice cabinet." The focus of the business model is on the home bartender, hence the limited offerings. Simeon explained, "Many other bitters companies approach the concept of bitters from a bartender mentality. And that's wonderful. Without that, we wouldn't have a lot of the great flavors that are out there. But our idea was to only make the two most useful, most versatile, most user-friendly flavors."

While the types of bitters made by Hella Bitter are familiar, the flavor profiles are unique. The citrus bitters aim for flavor, full spectrum as possible, for use in pretty much any cocktail that calls for citrus bitters. They use peels of orange, grapefruit, lemon, and lime, rounding out the citrus with ginger and whole spices, including cardamom, coriander, and black pepper.

The aromatic uses three bittering agents that help the flavor cut through the baking spices and the sugar of cocktails: angelica, gentian, and wormwood. There are still plenty of baking spices here including cinnamon, allspice, cloves, and star anise. Caraway seeds and black pepper add an earthy roundness.

Tobin Ludwig contrasted the two styles with a fun metaphor. "The aromatic bitters are like Christmas dinner at your Grandma's house; the citrus is more like summer in India."

PICKETT'S GINGER SYRUP

Jim Pickett has been at the New York City beer, wine, and spirits forefront for nearly two decades. His latest venture, with his brothers Bob and Matt, is Pickett's Ginger Beer Syrup. Jim's backstory: "I've always been kind of a ginger beer geek. And I tend to like mine a little bit spicier, because I like stronger spirits—anejo rum, bourbon, rye."

Where did the idea to make ginger syrup come from? "I also have another company that's a distribution company, and we were carrying a bottled ginger beer, but the supplier was very unreliable. My customers were getting frustrated, and one of my salespeople said, 'Why don't you just make your own ginger beer?'"

The logistics of bottling were intimidating and bad for the environment, so Jim came up with another solution. "I thought to myself, 'What about concentrated syrup for use with seltzer or a Soda Stream™?' I wanted to do it as natural as possible, cane sugar, natural ginger, and spices. I also wanted to target bars with soda guns, so we make a five-gallon bag-in-the-box version for them."

The Pickett Brothers Beverage office is here in Brooklyn, though the syrup itself is made Upstate and in Summit, Jersey. But Pickett gives full credit to Brooklyn. "The idea for the ginger beer was all formulated here. If I didn't have the experience I had in this borough, the product wouldn't be alive right now. Spending time in Brooklyn—I probably spend at least 50 percent of my work time out here—has pushed me to do this. Brooklyn is heavily influential on what I do on a daily basis."

TABLE No. 1
TRUE PERCENT PROOF

Hydrometer reading	Temperature °F.									
	1°	2°	3°	4°	5°	6°	7°	8°	9°	10°
76	101.9									
77	102.8	101.5								
78	103.6	102.4	101.0							
79	104.5	103.2	101.9							
80	105.3	104.1	102.7	100.6						
81	106.2	104.9	103.6	101.5						
82	107.0	105.8	104.4	102.3	100.2					
83	107.9	106.6	105.3	103.2	101.1					
84	108.7	107.5	106.2	104.0	101.9	99.8				
85	109.6	108.3	107.0	104.9	102.8	100.7				
86	110.4	109.2	107.9	105.8	103.6	101.5	99.4			
87	111.2	110.0	108.8	106.6	104.5	102.4	100.3	98.9		
88	112.1	110.8					101.2	99.8	98.5	
89	112.9	111.7							99.4	98.1
90	113.8	112.5							100.3	99.0

The financial collapse of 2008 was instrumental in the start of Steven DeAngelo's new career as a ginmaker. "I was working on Wall Street as an interdealer broker for a company named ICAP. Things were going well until Lehman Brothers declared bankruptcy in 2008. That was the beginning of the end. I never lost my job but things really went downhill and it started being a job that I hated." ▶

Of all the various options available for DeAngelo's next move, he chose opening a distillery. "It was the most interesting, yet most challenging idea I had. I read pretty much every piece of literature I could about distillation over the next several months, went out to Arizona to go to a distillation course, and decided I was going to dive into a business plan."

With a business plan in hand and full of determination, he still had a long road ahead of him. "It took almost three years to get up and running. I was trying to raise money in the worst economy since the Great Depression, without having a product, a name, a location, any experience, or pretty much anything else. So needless to say, nobody was interested in investing."

What he did have was a vision and a great idea—and a brother who saw the potential in what he was doing. "My brother, Phillip, said, 'Screw them. We'll fund it ourselves,' which turned out to be a great decision.

"After spending eight months looking for a location, followed by many months more of going through an absurd amount of paperwork, filing, and licensing for zoning and liquor licenses, we finally produced and sold our first bottle of gin on February 20, 2012, to Brooklyn Wine Exchange. A little less than two years later, we were available in about 900 locations in the tri-state area.

"The Greenhook American Dry Gin received rave reviews upon arrival, with the *Wall Street Journal* calling it 'one of the boldest, most interesting gins out there.'"

In June of 2012, the DeAngelos launched the world's first and only beach plum gin liqueur. DeAngelo explained the spirit's origin. "I was actually trying to do a traditional sloe gin when we launched, but had no idea that sloes don't actually grow in the U.S. I had contracted a farmer from England to ship me sloes, but he was actually a dairy farmer and didn't have the post-harvesting equipment to chill the fruit down to preserve it for shipping. There was a good chance the fruit would spoil before it got through customs. So I scrapped that idea."

Shortly thereafter came a revelation that yielded a more interesting product than if his Plan A had worked—a locavore version of a sloe gin. "I knew what beach plums were already, but later learned that they are very close relatives of the sloe. So I thought it would be a cool idea to make a New York ode to a sloe gin using beach plums, which only grow on the Atlantic coast from Virginia up to Maine." It was a brilliant bit of ingenuity.

His initial success and the promise of a bright future allowed DeAngelo to expand his operation a bit. In July 2012, he hired Joe O'Sullivan, a former distiller at the prestigious Clear Creek Distillery in Oregon. Steven told us, "I heard about him while having a drink at The Lodge in Williamsburg and complaining to the bartender there about how it was getting impossible to handle all of the distilling, sales, and deliveries on my own. She mentioned a friend of a friend who was in New York who used to work at Clear Creek." In July, 2013, Joe took over as head distiller, allowing DeAngelo more time to focus on sales and supporting his products. With 10 years' experience distilling already under his belt, Joe instantly became the veteran of the nascent Brooklyn distilling scene.

But where did DeAngelo develop his love of gin? "No one in our family has a history related to booze, except for maybe overindulging in it," he joked. "I was always a huge gin drinker, but mostly I drank Tanqueray and tonics and Martinis. To be honest, I was never really that big into interesting cocktails until recently, when all of these great bartenders started doing all of these amazing things."

What does the future hold for Greenhook Ginsmiths? Patience. DeAngelo's products can now be found throughout the tri-state area, plus Washington, D.C., Nashville, and even far off Singapore. But in the coming year, he is remaining resolute in maintaining his focus on succeeding fully in the New York market.

Because of his background in the finance world, DeAngelo understands full well the importance of careful, calculated business decisions to grow a strong company, especially

> "My brother, Phillip, said, 'Screw them. We'll fund it ourselves,' which turned out to be a great decision."
>
> —STEVEN DEANGELO

in an emerging field like craft spirits and an unforgiving market like New York City.

DeAngelo said, "I think that like anything else—products, services, or people—things that come out of New York have to be the best, because if you're not up to snuff you'll just go out of business, plain and simple."

So DeAngelo has slowly built a local, successful brand. Greenhook American Dry Gin received 94 points from *Tasting Panel* magazine, and his packaging received a gold medal at the prestigious Pentawards for Design. Gerard Caron, president of the Pentawards International Jury, said, "The work around Greenhook Ginsmiths is elegant and subtle, with a perfect blend of strong and original branding." That could also be said of what's in the bottle.

THE PROCESS

DeAngelo employs a rarely used method of vacuum distillation to capture the essence of his chosen botanicals. "For most of us in this young industry in this part of the country, the art of distillation has been very much self-taught. At the time I was planning my company, it wasn't like I could just go apprentice at a distillery for a few years to learn the craft."

So DeAngelo did what any good autodidact would do. He turned to books. "Most of what I learned was through careful study of distillation texts written in the 19th and 20th centuries." A lot of the material he found centered on perfume distillation. "I learned that French perfume makers started utilizing vacuum distillation because perfumes at the time were distilled almost purely from flowers, and flowers are extremely delicate and sensitive to heat, so traditional distillation temperatures destroyed their aromas and made the process extremely costly and inefficient. Vacuum distillation allowed them to remove air pressure from their stills, which in turn allowed them to distill at low temperatures, because boiling points vary at different air pressures." This method of low temperature distillation allows DeAngelo to keep the delicate aromas instead of steaming them off with excessive heat.

"Vacuum distillation really allows the aromas and flavors in my gin to stay intact; it creates a gin with very lifted flavors and aromas, and some lovely nuances that don't exist in other gins."

ALASKA COCKTAIL

2 1/2 oz GG American Dry Gin
3/4 oz yellow Chartreuse
dash orange bitters (optional)

Stir and strain into a cocktail glass.
Serve with a twist.

When we spoke to Steven DeAngelo about cocktails, he told us how much he likes the classics, and he singled out the Negroni as a drink that gets a lot of flavor out of very few base ingredients. The Alaska Cocktail is another drink in this mold that DeAngelo admires. "It's been a new winter favorite for me. It's a two-ingredient drink and my gin works really well in it." We agree. This gin is a local update on the classic London style. It's dry, clean, and infused with juniper throughout, but the nose is much more floral than a typical London dry gin. Chamomile and elderflower give it a very sweet, pleasant aroma, which muffles the Tuscan juniper a bit. The background cinnamon and citrus notes in the body pair very well with the herb, spice, and honey notes in the yellow Chartreuse. But don't take our word for it: put down the book and go mix one up for yourself.

RIVER'S EDGE

2 oz GG American Dry Gin
3/4 oz lemon juice
1/2 oz Dried Lavender Syrup

Combine the ingredients, shake, and strain into a high-ball glass on the rocks. Garnish with a fresh lemon twist.

DRIED LAVENDER SYRUP

16 oz water
16 oz sugar
3–4 tbsp dried lavender

Place all ingredients in a pan over medium heat and bring to just under a boil. Remove from heat and let steep for 30 minutes. Strain and bottle. It should keep in the refrigerator for about two months.

The first time we walked into Lavender Lake, we were thrilled to see many local bottles standing prominently on the top shelf, and the cocktail menu name-checked a few local spirits on its well-thought-out menu. This is one of our favorites.

This drink is rooted in a vodka-based Lavender Ice Tea that had been on the menu when Lavender Lake opened. The problem was that it wasn't quite to bar manager Thomas Callihan's taste. So he set out to use the key ingredient—lavender—and make something he liked better. As he told us, "Lavender, lemon, and gin go super well together, always." Again, the floral qualities of the Greenhook Gin are on display here. Elderflower and chamomile sing in three-part harmony with the lavender.

ROYAL ASCOT COCKTAIL

5 oz GG American Dry Gin

5 oz sweet vermouth

2 1/2 oz triple sec

1 1/4 quarts sparkling lemonade

1 1/4 quarts ginger ale

10 mint sprigs

1/2 cucumber, thinly sliced

1/2 orange, thinly sliced

1/2 lemon, thinly sliced

Combine all ingredients in a pitcher filled with ice. Pour into glasses, ice-filled or not, as you prefer, garnished with more cucumber, orange and lemon slices, and mint.

NOTE: When in season, strawberries make an excellent addition to this drink.

Obviously, you can feel free to use any lemonade and ginger ale you want, but for best results, try the following:

FOR THE LEMONADE

1 cup simple syrup (*see page 11*)

Juice of 10 lemons

4 cups chilled, sparkling water

Combine and stir.

FOR THE GINGER ALE

1 cup ginger syrup, preferably Morris Kitchen's or Pickett's, or even homemade

4 cups chilled, sparking water

Combine and stir.

When Pete graduated from New York University back in 1994, he took a trip to Europe. The trip was largely focused on drinking beer. In fact, it was like a graduate course in the brewing styles of Europe. There were many trips to pubs across London, learning what real ale was. Belgium was like beer heaven, and plans were rearranged so he could spend extra time there, doing day trips from Brussels to Antwerp and Bruges, learning about bière de garde and saison, as well as various lambics and geuze.

The whole six weeks away, he had but one cocktail—a Pimm's cup at a students' party in Edinburgh, Scotland. It was pretty basic: Pimm's No 1, lots of fruit, too much 7-UP, but the flavors stayed with him, even though he didn't have another for nearly two decades when he had a pint of the stuff at Royal Ascot in 2010.

It was too long between drinks. We recently got this recipe for a "homemade" version of Pimm's from Building On Bond, rebranded here as the Royal Ascot Cocktail. It's simple as can be. And the touch of a little homemade ginger-lemon soda takes it to the next level.

FEMME FATALE

2 oz GG American Dry Gin

3/4 oz grapefruit juice

1/2 oz lime juice

1/2 oz simple syrup (*see page 11*)

1 tsp St. Germain

2 leaves of basil (*one for the shake, one for garnish*)

white of 1 egg (*or 3/4 oz carton egg white*)

splash of Angostura aromatic bitters for the meringue

First, do a dry shake of all the ingredients
(no ice) to set up the emulsion.

Then add ice and shake really hard. The shaker
should start to frost over and stick to your hand
when it's been shaken enough. Strain into a rocks
glass with an ounce of soda in the bottom. If it got
shaken enough, it should froth up to the top.

Garnish with an additional basil leaf and
up to three lines of bitters right in the meringue.

When we sat down to write this book, we didn't intentionally set out to include every common herb, but it kind of worked out that way. This cocktail was devised by our pal Will Noland at Sidecar. It's a great-looking drink and a conversation starter even: "Ooh, what is that?"

It isn't always made with the Greenhook American Dry Gin, but the combination works really well, with the cocktail ingredients and the gin sharing certain herbal and floral qualities. Plus, who can resist a drink named after a Velvet Underground song?

Noland's quote about the Femme Fatale gives you some insight into his personality, at once humble and confident in his craft: "I don't want to sound like a dick, but it's a damn good drink."

BRAMBLE

2 oz GG American Dry Gin
1 oz fresh lemon juice
1/2 oz simple syrup (*see page 11*)
1/2 oz Will's Blackberry Liqueur

In a shaker filled with ice, shake together the gin, lemon, and syrup, and strain into a rocks glass filled above the brim with ice (preferably crushed ice). Pour the crème over the top and garnish with a blackberry and lemon twist. A short straw or extra blackberry wouldn't hurt.

The Bramble is the perfect cocktail vehicle for consuming good blackberries when they're in season. And when they're not in season, Will Noland's housemade blackberry liqueur (crème de mure) still gives the crowd at Sidecar a chance to enjoy the freshness of summer fruit throughout the year, even in the dead of winter. In a drink like this, Greenhook Gin shines—the flavor of the blackberries is intensified by the spicy cinnamon prominent in this gin.

WILL'S BLACKBERRY LIQUEUR

1 pint blackberries
1 pint sugar
1 pint vodka

Place blackberries and sugar in a blender ("No fancy ass blender required," Noland assured us). To ensure all the fruit gets blended, add only a few ounces of vodka at first and blend. Then add the rest of the vodka and blend again. A lot of tasting is required, as the recipe will vary slightly depending on how sweet the fruit is. You'll want to add sugar for sweetness and/or more vodka for viscosity, depending on your taste. When you've achieved the desired sweetness and consistency, strain and bottle.

Our advice is to do what Noland did—start by buying a bottle of artisan-made crème de mure so you can get a good idea of what the final product is supposed to taste like, then experiment and tweak accordingly so your version suits your tastes.

This technique will work as a way to preserve other fruits as well, though the proportions may vary somewhat. As Noland told us, "If I'm making one with rhubarb, I'm going to use a lot more rhubarb if I want to get the same amount of flavor because the fruit flavor is a lot less forward."

"I always liked brambles but wanted a fresher character and something less sweet," Will Noland says. He did his homework and sought out the best artisanal bottle of crème de mure he could find. "I saw what was in it and said, 'I'm going to make this myself.'"

LAVENDER LAKE

(Gowanus)

In the mid-90s, Pete was walking from Carroll Gardens to Park Slope and saw what looked like your average New York City street person—ripped flannel shirt, shopping cart—ranting and raving, "I have the seen the future!" he railed to the wind. "Gowanus!"

Upon hearing the word "Gowanus," Pete got interested and slowed his double-time fast city walk to more of a quick amble. He wanted to hear what the Fisher King had to say about the famously filthy canal that smelled like low tide morning, noon, and night. "Some day," he preached, "It will be like Venice! Restaurants! Hotels! Gondolas!"

Nearly two decades later, this vision hasn't exactly come to fruition, but it's also not quite as crazy as it once seemed. There are businesses more or less on the banks of the Gowanus, as evidenced by the very existence of Lavender Lake, whose name is an ironic allusion to the Gowanus, the Lavender Lake in question. So even if there aren't any gondolas, this is still a pretty good start.

Lavender Lake is a sweet spot with a minimal, modern décor that still manages to have a warm feel, with its brick and wooden beams. The outdoor patio is huge—1600 square feet—and is obviously ideal for warmer weather, when Lavender Lake hosts special events, including grill specials and movie nights. The food has a bit of a West Coast vibe, with fresh fish and smart takes on pub classics. And the drinks are the real deal, very much focused on local/artisanal spirits.

Part of why Conrad Oliver and Thomas Callihan from Lavender Lake like supporting the local spirit makers is that they can relate to them. Callihan told us, "We're a small business, just like these guys are; we support little brother, not big brother, because that's what we are." Oliver added, "And there are definitely layers to it, too. We have the opportunity to build a relationship on a more personal level. I see nothing wrong with Ketel One or Belvedere, but we replaced those guys with Industry Standard vodka from Industry City Distillery, which is two miles down. Those guys support us and we're so happy to reciprocate."

FRENCH 75

2 oz GG American Dry Gin

4 oz chilled Champagne to fill a flute

1/4 oz simple syrup (*see page 11*)

1/4 oz lemon juice

lemon twist, ideally a horse's
neck-type twist (done with a
channel knife) like in the photo

*Build the cocktail in a Champagne flute or
coupe. Garnish with the lemon twist.*

There are some drinks in this book for which we—or the bartender who created the cocktail—labored to find something specific in the spirit of choice that makes it uniquely appropriate to the mix of ingredients in the recipe. This is not one of those.

The French 75 is a classic Champagne cocktail preparation that would work quite well with just about any spirit in this book, including the whiskeys. But the classic flavor profile of the Greenhook American Dry Gin does well in this preparation. For an easy variation, pour in a little quarter ounce of the Plum Gin Liqueur as well.

WINTER CHERRY SMASH

1 1/2 oz GG American Dry Gin

1/2 oz maraschino liqueur

2 oz Boiron Black Cherry Purée

1 oz lemon juice

2 dashes Angostura aromatic bitters

*Combine ingredients and shake well, then add
a splash of soda to the shaker. Strain over
fresh ice and garnish with a lemon twist.*

Henry Lopez of Lucey's Lounge makes good drinks—really good drinks. In fact, this one even won an award from a website called World's Best Bars (worldsbestbars.com). Despite the accolades, Lopez considers himself a bartender, not a mixologist. "I'm convinced that a mixologist is what bartenders need to call themselves after the age of 35 to make themselves feel better about what they're doing for a living."

There is no better place in Brooklyn than Lucey's Lounge when it comes to combining the concepts of the neighborhood bar and the cocktail bar. Lopez went on to explain what might be considered the Lucey's Lounge mission statement: "We like people, we like to tell stories. We like to make great drinks with the best ingredients possible, and not make any big whoop about it. And I'll talk your damn ear off."

PLUM GIN SILVER FIZZ

1 oz GG American Dry Gin
1 oz GG Beach Plum Gin Liqueur
3/4 oz fresh lemon juice
3/4 oz simple syrup (*see page 11*)
1 small egg white
dash Peychaud's Bitters

Dry shake the ingredients.
Add ice and shake again vigorously.
Strain and top with soda. As an
alternative, omit the bitters from
the shake and add a few drops to the
meringue.

Well, sloe gin fizz works mighty fast, when you drink it by the pitcher and not by the glass.

This drink is an updated American version of the cocktail Loretta Lynn sang about in "Portland, Oregon," though it uses Greenhook Beach Plum Gin in place of the sloe gin (you'll find the specifics of why this substitution works so well earlier in this chapter).

In England, sloe gins are made from actual sloes, a stone fruit that comes from the Blackthorn tree. There is also a DIY culture associated with the liqueur, with people around the English countryside frequently macerating foraged sloes in high-proof gin, then adding sugar. This distinctly English product is surrounded by a culture of national pride—to the point where each January, a Grand Master of the Sloes is declared at a competition in Dorstone, Herefordshire.

In 19th-century America, fizz drinks had a different purpose—they were hangover cures. Taken in a small glass and topped with fizzy water, they were downed as a way to get the ball rolling on a new day.

A few of the properly made sloe gins have made it to our shores in recent years, but for the longest time, America's answer to sloe gin was a sickly sweet, artificially colored and flavored concoction that found its home in drinks like the Alabama Slammer, the type of drink Tom Cruise's character rhapsodized about in the movie *Cocktail*.

For the gin we're using here, whole beach plums from Long Island are macerated in Greenhook Ginsmiths American Dry Gin for seven months. A little sugar is added and the proof is brought down to 60. The result is a glorious magenta liquor, lightly sweet and fruity, redolent of juniper and spicy gin, but slightly tannic in the body due to the inclusion of the plum pits.

As for the Loretta Lynn quote above, we have yet to see this version of a sloe gin fizz served by the pitcher, but on one hot September afternoon at Belmont Park (New York's largest outdoor alcohol-friendly picnic area), Steve DeAngelo did bring along a pitcher of Plum Gin Collins. Same recipe as below, just omit the egg and bitters float and serve in a taller glass over ice.

OLD-FASHIONED TOM

2 oz GG Old Tom Gin

1/4 oz simple syrup (*see page 11*)

dash orange bitters

dash Angostura aromatic bitters

1 blood orange slice (*or orange slice*)

1 homemade cocktail cherry (*see page 159*)

In a shaker mostly filled with ice, combine the bitters, syrup, and Old Tom, and stir. Strain into a rocks glass containing ice, and garnish with an orange slice and homemade cocktail cherry.

Note: Boelte prefers a turbinado sugar, like Demerara, for his syrups.

For Greenhook Ginsmiths Old Tom Gin, Steven DeAngelo reached out to the local cocktail community. He tapped Maxwell Britten of Maison Premiere and Damon Boelte of Prime Meats to help him construct a flavor profile. Boelte described what they brought to the process. "One of the things that Max and I really wanted to do was to play with perceived sweetness. We didn't want to overly sweeten it with sugar; we wanted natural sweetness. We wanted to play with flavors like vanilla and lavender. We were looking at different casks for finishing. We wanted to use the caramelized sugars from a bourbon barrel plus the residual bourbon itself, and that's where you get the vanilla from, and then we finish it in sherry casks so you get a little bit of sweetness from that as well."

Old Tom is an old friend in the gin family—one that has been away a while—and an essential ingredient in classic 19th-century cocktails like the Martinez and the Tuxedo. The current boom in local distilling brings with it the resurrection of old, favorite styles of spirits that were prevalent in the late 19th century's "golden age" of cocktails. While there is no extant clear definition of what makes an Old Tom, it is always expected to be sweeter than London dry style gin, and often it is aged in wood barrels, which add color and flavor. The sweetness of the gin allows you to build a cocktail with less added sugar in the form of syrups, liqueurs, etc.

Greenhook Ginsmiths produce their Old Tom Gin from a mostly typical gin botanical base, which is then lightly sweetened with the addition of un-aged corn whiskey. Then it is aged for a full year in used Jim Beam barrels before being finished in Oloroso sherry casks and bottled at 100.3 proof.

This classic cocktail is a fine way to feature Greenhook Ginsmiths Old Tom Gin.

HOMEMADE COCKTAIL CHERRIES

GUEST RECIPE BY TOBY CECCHINI

You need to get dark sour cherries for this recipe. Sweet cherries like Bing and Rainier are just too flabby and don't give enough bite or aroma when preserved. I usually use either Balatons or Sure-Fires, and sometimes get both. These typically ripen around the middle of July, but the season can vary. They are difficult to find, so you may have to do some digging around your locality or online. I get mine from Upstate New York, but Michigan, Wisconsin, and Oregon are good producers as well. They take a good six months to a solid year to become what they're supposed to, and the best part might be the resulting elixir, a 1/4 ounce of which you can then add to your Manhattan: amazing.

Into a cleaned half-gallon (32 oz), wide-mouth Mason jar, put:

2 bourbon vanilla beans, split lengthwise

4 1/2–5 lbs dark sour cherries per jar, rinsed beforehand in cold water, and with detritus sorted out. You can leave on the stems and some of the leaves, and DO NOT pit them; all of these elements add essential bitter (the pits) and vegetal flavors to the end result, and the cherries maintain their form much better whole.

1 quart spirit mixture

Make the mixture of:

1/3 quart kirsch; I use Maraska Extra Dry Kirsch (90 proof)

1/3 quart rich syrup (*see page 11*). A Demerara or turbinado sugar adds some heft to the resulting elixir.

1/3 quart overproof spirit, at least 100 proof. You can use vodka. I had my best result ever with J.M Blanc rhum agricole one year and continue now to use combinations of white, overproof rums.

That's it. No heat. The spirit will sterilize and preserve of its own accord. Give the cherries at least six months and really a year at best, to become properly preserved and imbue the elixir with their flavor. They last indefinitely; I'm still happily using batches four and five years old.

EAST MEETS EAST

1 oz GG Old Tom Gin

1 oz blended Scotch (*Boelte mentioned Pig's Nose and Compass Box as brands he likes for this*)

3/4 oz dry vermouth

3/4 oz sweet vermouth

dash orange bitters

dash Angostura aromatic bitters

flamed orange peel

Place the first six ingredients in a mixing glass mostly filled with ice. Stir very well. Meanwhile, flame an orange peel under a chilled cocktail glass (see note).

FLAMING THE PEEL

Here's Boelte on the flamed orange peel for East Meets East: "Rather than use matches, I like to light a wooden coffee stirrer. I heat up the peel a little bit first and I flame it underneath the glass, not in it. For this drink, I'm not looking to get the singed oils in the cocktail. I just want the aromatics from the flamed peel."

Damon Boelte of Prime Meats is one of the most recognizable bartenders in Brooklyn. We were thrilled when he invited us to come by Prime Meats and make some drinks with the Greenhook Old Tom Gin. Boelte broke down this unusual cocktail for us: "This is a hybrid cocktail of a Rob Roy and a classic Martini. It's basically a 50/50 combination of each, just heavier on the vermouth. Julie Reiner called it 'strong and strange,' but I'm pretty sure she meant it as a compliment."

It is both strong and strange, but it absolutely works, especially with the Old Tom in place of the London dry gin Boelte was using previously. "The botanical blend in the Old Tom pairs well with the Scotch. And since the Old Tom is barrel aged, it's not too much of a leap for the two base spirits to blend together well."

STRONG ISLAND

1 oz GG Beach Plum Gin Liqueur

1 oz Dorothy Parker Gin*

1/2 oz lemon juice

1/2 oz simple syrup (*see page* 11)

dash Regans' Orange Bitters No. 6

sparkling wine

Place all ingredients in a cocktail shaker mostly filled with ice, shake, then strain into an ice-filled collins glass. Finish with a float of sparkling wine and an orange twist.

*(Obviously the GG American Dry Gin would work great too, but this is how they make it at Marlow and Sons.)

Andrew Tarlow is one of the key figures in the Brooklyn food scene. When his restaurant Diner opened in 1998, it put Brooklyn on the map in terms of contemporary dining. Marlow and Sons, next door to Diner, is a great destination as well for coffee and pastry in the daytime, cocktails and snacks at happy hour, and Mediterranean-influenced cuisine at night.

"We are huge supporters of all the various spirits being made in Brooklyn," Tarlow said. "We have this policy where we try not to buy food from strangers, and we try not to buy wine from strangers. In a similar way, we'd rather know the people who make our vodka and gin. We'd rather support the individual owner rather than the large liquor conglomerate."

Tarlow explained the philosophy behind his cocktail menu. "We try and stay seasonal, and we focus on how you begin a dining experience and how you might want to end a dinner."

With that quote in mind, this cocktail, whose name is an irresistible pun, would make a lovely aperitivo on a warm summer evening. A word of warning: as bartender Bree Nichols said, "You have to be careful with these because you drink a few of them before you realize how wasted you've become."

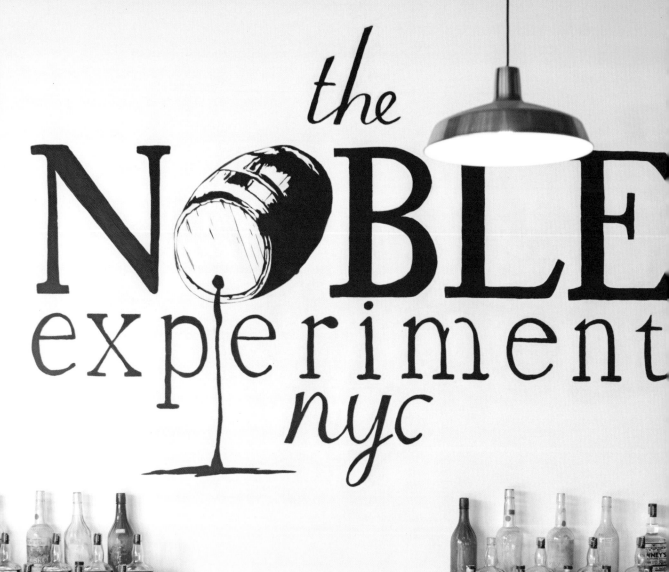

In December of 2011, Bridget Firtle made a life-altering decision. She left behind her lucrative hedge fund job in Syosset, New York, gave up her tony Tribeca loft, and moved back into her childhood bedroom in Rockaway Beach with her parents. Why on Earth would someone do such a thing? ▶

For Firtle, it had a lot to do with the people she met in the course of her everyday work as a securities analyst for global alcoholic beverages at Knott Partners. She had been interacting with brew masters, distillers, and wine makers on a regular basis, and the constant exposure to their world awoke something within her.

As 2012 dawned, she assumed a new title, CEO & founder of The Noble Experiment, NYC. Bridget Firtle was going to own and operate her own distillery and reintroduce people to the all-but-forgotten joys of premium rum. Not only was she the only female distiller on the local scene but she was also one of the youngest. Many people told her she was crazy. Firtle was undaunted, and for that we can be grateful.

We recently took a tour through her Bushwick distillery and had a chance to talk with her about her life, the history of distilling in New York, and her decision to make rum. She was expansive on all three subjects, particularly the latter. "I love all types of rum. I think rum is cool. I think a lot of people don't think rum is cool, and I want to teach people that rum is awesome. There's such a rich history with rum distilling in this country; we haven't distilled rum here in a really long time. I thought it was about time we bring rum distillation back to the Northeast, more specifically New York."

It is no accident that Firtle has taken up the stewardship of the rich history and legacy of rum distilling in New York City. One might even say that destiny played a hand in her unusual vocation. Born in Brooklyn Hospital on DeKalb Avenue and educated in the New York City public school system, she went on to earn a bachelor's degree in business at McGill University in Montreal and an MBA in finance at SUNY Binghamton. Her family often uses the term "feisty" to describe her, and we can see why. The sight of the former business major behind the wheel of her forklift, deftly hoisting 700-pound barrels of molasses into the air, is impressive, to say the least. From distilling to bottling to sales and promotions, Firtle is a one-woman show.

Booze business acumen runs strong in her family. Her grandfather owned and operated a "beer-and-a-shot bar" called Coughlin's on Flatbush Avenue from the 1950s through the 1970s. And as if that weren't enough of a connection to the world of alcoholic beverages, the basement of the house she grew up in housed a speakeasy during Prohibition.

Yes, Prohibition. Officially the National Prohibition Act of 1920, also known informally as The Volstead Act (and often referred to as "The Noble Experiment," which, of course, would become the ironic name of Firtle's distillery 92 years later). Once it was passed, drinking was against the law; for the next 13 years, people who continued to drink were forced to "speak easy" because they might get caught. With the clarity of hindsight, it is easy to lament the ensuing rise of organized crime, the billions of dollars in lost revenue, and the death of about a thousand New York State distilleries, but Firtle takes a different approach, preferring instead to celebrate the resiliency of the human spirit. For during this dark period in American history, thanks to the efforts of some enterprising, albeit nefarious, characters, many people continued to drink.

Firtle's explanation for why she chose to name her flagship product after notorious gangster Owney Madden is every bit as rich as the distillation which bears its name. Owen "The Killer" Madden, also known as "Owney," was a Prohibition-era rum-runner, a Hell's Kitchen gang leader, a bootlegger, and a speakeasy operator. Owney smuggled rum from the Caribbean to New York via the waters off Rockaway Beach, where he owned an estate—a mere stone's throw from what would become Firtle's birthplace decades later. The cargo ships would anchor 12 miles offshore, in international waters where rum was still legal, at a spot that came to be known as "Rum Row." Smugglers would then row out to meet the ships in the dead of night in order to ferry their contraband back to shore. Madden's illicit activities, so the legend goes, helped rum to regain some of the popularity it had once enjoyed as America's favorite booze during the Colonial era.

Firtle explained why she chose Brooklyn. "I never wanted to capitalize on the hype behind Brooklyn right now. I happen to be from New York. Both sides of my family are from Brooklyn; it's the borough I know best. It's also the borough that actually has space enough that you can do something like this."

And then there is the matter of Owney's place in the resurgence of the local cocktail culture. Here again, Firtle proves herself to be a practical thinker. "I try to go and spend money in the places that carry my product, chat the bartenders up, and meet the people who are enjoying it." As is the case with many of our featured local distillers, Firtle's admitted personal enjoyment of cocktails takes a backseat to her professional love of, and advocacy for, all things Owney's. "I'm a better distiller than I am a bartender," she says. When pressed on the cocktail issue, her historical perspective comes shining through in her discussion of the classic Daiquiri. "Every cocktail program should have a Daiquiri, or some variation of it." She also expresses fondness and enthusiasm for a "good Mai Tai." You'll find recipes for both of those later in this chapter.

THE PROCESS

What is it that makes Owney's Original New York City Rum so special? What sets it apart from the competition? Firtle's face lit up as she described her baby in a bottle. "There are a number of things that make Owney's unique. First of all, the ingredients: It's made from New York City tap water that's been filtered. I think it's the best water in the world. As a native New Yorker, maybe I'm biased, but from a scientific standpoint, the chemical composition is perfect for distilling. Also, I use proprietary yeast. And lastly, I use an all-natural, non-GMO, top-grade sugar cane molasses that comes from small, independent sugar cane farms in Florida and Louisiana, so it's 100 percent domestic."

It would have been easy for her to replicate a clear commercial rum, made from blackstrap molasses, and pass it off as special just because it's a local product. But she wanted to be better. Her love affair with rums naturally led her to the much admired *rhum agricole* style of production found mostly in former French colonies, which uses pure cane juice in place of molasses. Cane juice is sugar rich and has the distinct flavor of its source—known as *terroir* in the wine world. Rhum agricole can also be grassy and assertive—positive qualities in a spirit category that's become known for its general lack of flavor. But pure cane juice can be very temperamental and difficult to ship, so Firtle decided to create a rum that, while made from domestic molasses, reflected everything that was desirable about the flavor profile of *rhum agricole*.

She starts with the top grade of molasses, which contains 80% sugar (as opposed to the 10% of blackstrap), because it hasn't had all of the sugars refined out of it. This leaves the molasses flavorful and aromatic and full of potential. The high sugar content enables her to ferment the molasses similarly to how cane juice is distilled. She explained to us, "I do five-day-long cold fermentations as opposed to the typical 48 hours for a blackstrap molasses. I keep the yeast alive, happy, and eating those sugars. Because of all of the sugar in the molasses, there is a lot of clean alcohol available in the wash to extract during distillation." This process also leads to a lot of desirable flavor by-products (esters) that help give Owney's its unique flavor.

Next, she distills the fermented wash to 82.5% alcohol (165 proof). Again, this is a cross between the two styles of rum, where white rum (from Spanish-speaking countries) is typically distilled to upwards of a flavorless 95%, and *rhum agricole* is distilled to maximize natural flavor at as low as 70%. Firtle explained, "I'm meeting in the middle at 82.5%. It's where I think it's got a perfect balance: not too much bite, and a lot of flavor."

It's hard to mistake a cocktail made from Owney's rum—and that's a good thing. Owney's is often described by an adjective that is only positive in music and rum: *funky*. Rum drinkers look for a little bit of funk in their rum to help it stand out in a great cocktail, rather than just blending in the background and adding sweetness and viscosity to what might already be a sweet, heavy drink. That funk is a sulfurous, tangible tang that usually results from the distillation of sugar cane juice. Since Owney's is distilled in a hybrid process, it has the best qualities of a molasses-based rum and rhum agricole...just funky enough.

MIDTOWN BOOTLEGGER

2 oz Owney's rum
3/4 oz lime juice
1/2 oz Cinnabark Syrup
1/2 oz Licor 43
1 pinch grated nutmeg

Place the first four ingredients in a cocktail shaker mostly filled with ice. Shake well and fine-strain into a frozen coupe. Garnish with grated nutmeg.

CINNABARK SYRUP

This Cinnabark Syrup recipe comes from the famous Clover Club, Nate Dumas's home before the Shanty. Bartender Tom Macy gave us the house recipe and his take on cinnabark versus cinnamon syrup. He tells us, "The two are kind of interchangeable. We use cinnabark at Clover because it's a little more flavorful, but it's harder to find, so cinnamon sticks work too."

1 quart sugar
1 quart water
2 cups cinnabark (or 10-12 cinnamon sticks)

Break up the cinnamon sticks into small shards, or crush the cinnabark. Combine the ingredients in a small pot and bring the mixture to a boil, then remove from heat and let sit overnight for best results. Strain and bottle.

Designed by Clover Club's head bartender, Tom Macy, the Midtown Bootlegger is his personal, seasonal riff on his favorite cocktail six months out of the year—and one that appears a lot in this chapter—the Daiquiri. He explains, "I was working on drinks for Clover Club's fall menu, so that meant I was looking for classic autumnal flavors. Instead of the traditional sugar, I decided to sweeten the drink with cinnamon syrup and Licor 43, which is a Spanish liqueur with forward notes of vanilla that paired perfectly with Owney's rum. I find the Owney's has a lovely rich and buttery finish that remains light enough to keep the drink true to its roots." Still, Tom felt there was an element missing. "It wasn't until I added the grated nutmeg, a very traditional ingredient in old rum punch recipes, that the cocktail really started to sing."

The drink's name is an obvious allusion to Owney Madden's infamy as a bootlegger. "I was initially going to call the drink Hell's Bootlegger, but it just seemed a little over the top."

ORGEAT WORKS

They call him Tiki Adam. He is one of the great characters on the New York/Brooklyn bar scene. Adam Kolesar developed a cocktail obsession after taking a rum punch class from Dale DeGroff. Kolesar shared some historical perspective. "The birth of Tiki is really about the post-Prohibition availability of rum. Rum was the cheapest and best spirit available post-Prohibition, and it was something that Don the Beachcomber knew a lot about from his travels. With his intimate knowledge of rum, it just made sense for him to do rum punches."

But Kolesar's mania for cocktails isn't rooted in the usual causes. "Drinking isn't about the drunk for me," he told *New York Times* food critic Frank Bruni in 2011. "It's about creating the taste profile of an era I romanticize and care about."

For him, that means knowing every detail of Tiki culture from its true origins, to its appropriation at Trader Vic's, to its renaissance in the last 10 years. "My approach to drinking is anthropological," he told us at his home bar, made out of an old Airstream. "I don't create drinks. I'm about researching and developing the taste profiles that Don the Beachcomber and Trader Vic developed in the origin of Tiki. I'm most interested in experiencing what our forefathers experienced when these drinks were conceived."

In 2013, Kolesar started Orgeat Works. Orgeat is an almond-based syrup that is a *sine qua non* in many Tiki classics, from the Mai Tai to the Scorpion Bowl. His signature product is a lightly toasted version of a classic orgeat, but he also makes a heavy-duty orgeat and a macadamia nut version. "I'm looking for products that will work well with all these amazing recipes," Kolesar said. "And I want to make syrups that cocktail lovers and Tiki geeks will really like and want to try."

MAI TAI

1 oz Owney's rum

1/2 oz Appleton Special Jamaica Rum

1/2 oz El Dorado 15 yr old rum

1/2 oz Orgeat Works Toasted Orgeat Syrup

1/2 oz Pierre Ferrand Dry Orange curaçao

1 oz fresh lime

Fill a rocks glass with crushed ice. Pour the Appleton's and El Dorado over the ice. Next, combine the Owney's rum, lime, Orgeat, and curaçao with plenty of ice in a shaker. Throw in the lime hull to scuff the remaining lime essence from the rind. Shake away and strain over the crushed ice. Garnish with a generous sprig of fresh mint.

The Mai Tai is the *Sergeant Pepper's Lonely Hearts Club Band* of Tiki drinks. In one bold move, Trader Vic changed and reinvigorated an entire genre of cocktails, if not a whole lifestyle, and the rest of the world has been trying to catch up ever since. We asked Adam Kolesar to describe the importance of a good orgeat (read: his) in a Mai Tai. Not surprisingly, he had a good answer:

"The working theory behind my orgeat was that in 1944, when Trader Vic invented the Mai Tai, the orgeat had some semblance of the taste of almond. The toasting, for me, is what helps bring the real essence of the almond forward. It's a way to get almond flavor prominent without resorting to anything artificial. It just makes sense in that drink that you should have an almond component with a bitter finish, rather than a perfume product, which is what commercial orgeat had become. That combination, along with the rum and the curaçao, is what makes a Mai Tai a Mai Tai. You don't have a Mai Tai without orgeat. You can't sub it out without destroying the drink's complexity.

"In this Mai Tai, Owney's rum provides the critical 'funk' element, traditionally expressed through an aged agricole rum. The Owney's plays well with the Appleton's Gold and El Dorado Demerara rums to create a unique Mai Tai that celebrates Trader Vic's original vision."

The result is spectacular.

THE RICHARDSON

(Williamsburg)

"A return to classics, across the board, is a good thing," Steven Spate, manager and senior bartender of the Richardson says. "That pretty much defines the idea behind our cocktail program. We're rooted in classics: classic sprits, classic drinks, and a conscious avoidance of any sort of nouveau approach to drinks—whether that means over garnishing or over muddling or overdoing it with house-made stuff. We make our own grenadine because it tastes better than any grenadine you can buy, but we don't need to make everything. We're not looking down on any of that stuff, it's just not what we do here. We'll continue to experiment and adapt, but we're also going to be rooted in the classics."

DAIQUIRI

2 oz Owney's rum
1/2 oz fresh lime juice
1/2 oz simple syrup (*see page 11*)

Pour into an ice-filled shaker, shake, and strain into a cocktail glass or coupe, preferably chilled.

When you take a tour of The Noble Experiment distillery, there is a question-and-answer segment at the end. People frequently ask Firtle how she likes to enjoy her rum. Her answer makes jaws drop to the floor—such is the state of the poor, beleaguered Daiquiri to many drinkers in America.

In reality, the Daiquiri is as classic and simple as it gets. It's the way most bartenders like to test-drive a new rum. But over time, the Daiquiri has morphed, eroded really, into something else.

"I would spend an hour on the tour lovingly describing my process and my rum, and then people would think I was telling them to dump it in a blender with ice, a box of sugar, and all these artificial flavors. But a real Daiquiri isn't like that at all. It's a template from which you can create this whole world of interesting flavors."

The basic Daiquiri recipe listed here comes from The Richardson. It's a fine cocktail as is—with no special DIY requirements or even bitters to add. You should feel free to change the proportions and/or substitute/add different elements to cater to your own tastes, but we think the Owney's acquits itself beautifully in this preparation.

MISS MILLIONAIRE

3/4 oz Owney's rum
3/4 oz Rothman & Winter Orchard Apricot Liqueur
3/4 oz Averell Damson Gin Liqueur
(*you could try GG Plum Gin Liqueur instead*)
1/2 oz lime juice
1/4 oz Homemade Grenadine

Place all ingredients in a shaker mostly filled with ice, and shake well. Strain into a cocktail glass.

HOMEMADE GRENADINE

Guillermo Bravo: "We make it by combining equal parts organic pomegranate juice and sugar brought to a simmer with constant stirring. Once all the sugar is combined, turn off the heat and let it cool. It ends up the right amount of bitter and sweet."

Manager and senior bartender Steven Spate of The Richardson is a big fan of Owney's in cocktails. "Owney's holds up great even in cocktails where you initially think it might not fit. The Millionaire is a classic example. Usually you'd use an overproof rum, but we wanted to try Owney's in there and it really works. You get a lot of the character of the molasses Firtle is using. It has so much character that even as a white rum, it can stand in in classic cocktails for a gold or dark rum."

THE BAD HABIT

1 1/2 oz Owney's rum

1/4 oz cocktail cherry syrup (*Blegen uses
syrup from Luxardo brand cherries*)

1/2 oz fresh squeezed grapefruit juice

3/4 oz fresh lime juice

2 tsp sweetened ginger purée*

*Place all ingredients in a shaker mostly filled
with ice. Shake vigorously to incorporate ginger.
Double strain over a grapefruit cube.*

TO MAKE GRAPEFRUIT CUBE:

*Cut away the sides of the grapefruit to leave you
with a cube. Reserve the sides for juicing. Place
grapefruit cubes on a parchment lined-tray and put
them in a freezer for two or three hours. Once
they are frozen, wrap individually with plastic
so they don't stick together while being stored.*

**Blegen uses a ginger purée from Perfect Purée
in Napa Valley but you could experiment
by using puréed ginger and sugar.*

This cocktail, designed by Alicia Blegen of Marietta, was inspired by an unlikely source: a donut.

"I was over at Pies and Thighs in Williamsburg, having lunch with a colleague, and I ordered a grapefruit-ginger donut, and it was so juicy and delicious. At the time, I knew I wanted to use Owney's on the menu. I love Bridget and her rum is so unique and great, and the Owney's doesn't mix quite the same as your typical light rum. It's a little rawer and reminds me almost of a tequila. I thought about a Daiquiri, then I thought about adding grapefruit and doing a Hemingway Daiquiri, and then I remembered that delicious donut from Pies and Thighs. I wanted to recreate that donut in an intoxicating form."

Another really cool and special thing about this drink is the garnish, which reflects Marietta's nose-to-tail, or in this case, core-to-rind ethos. Every part of the grapefruit is used: the zest gets shaken in drinks, the outer fruit gets juiced, and the big block in the center gets frozen for this drink. It's a treatment that world-renowned chef and ice enthusiast Grant Achatz would be proud of. But this drink is no mere gimmick—the flavors amplify one another perfectly, and it's an excellent application of the Owney's rum.

THE AIRMAN

2 oz Owney's rum
2 tsp créme de violette
3/4 oz fresh lime juice
3/4 oz simple syrup (*see page 11*)
1 drop rose water

Place all ingredients in a cocktail shaker mostly filled with ice, shake, and strain into a chilled cocktail coupe.

The Airman is a unique take on the Aviation cocktail, using rum as its base instead of gin. It gets its name from an English children's book that Dear Bushwick partner Julian Mohamed remembers from his childhood in Oxfordshire. He recently brought a copy back from his home-town, and it now sits on a shelf in the restaurant.

Darren Grenia explained this cocktail's origins to us on a recent visit. "I really like Bridget and the work she's done with Owney's. My initial thought with this one was a loose play on the Aviation cocktail, but I knew that this rum made an amazing Daiquiri, so I decided to morph the two together."

We've never seen a drink like it—and we mean that literally. Depending on the light, this cocktail can look purple, green, purplish-green, gray, blue, or some combination of all of the above.

DEAR BUSHWICK

(Bushwick)

Dear Bushwick is one of the coolest restaurants in Brooklyn, hands down. It's modeled after an English country inn as seen through the lens of 21st-century Brooklyn. The food has won awards and we can see why—the seemingly simple Eggs and Butter is a miracle of complexity, cooked with wine-barrel-smoked eggs, a horseradish compound butter, and green peppercorn salt.

But this is a book about cocktails, and Darren Grenia makes great ones. In fact, while the plans were not finalized as of this writing, one project for the near future includes the partners at Dear Bushwick opening a speakeasy of sorts next door cleverly called, Sincerely Yours. Darren says, "Our vision for it is to create an elegant, grand experience in Bushwick. A sort of cross between an opium den from the 1800s and a Victorian-era mansion with a focus on barrel-aged cocktails." He had us at "opium den." Whatever the team behind Dear Bushwick does next, we'll be watching with great interest.

MEADOW STREET GARDEN

2 oz Rosemary-Infused Owney's
splash of Solerno Blood Orange Liqueur
(*you could substitute another orange liqueur like curaçao or triple sec*)
a few small pieces of rosemary (*one large spring total*)
4 segments peeled tangerine
4 segments peeled blood orange

Muddle about half the rosemary and half the citrus segments in the bottom of a cocktail shaker with the orange liqueur. Add ice and rosemary-infused rum and shake vigorously. Strain over fresh ice in a glass with additional rosemary and fruit segments placed throughout. (see photo)

ROSEMARY-INFUSED OWNEY'S

Take one large 8–10 inch healthy sprig of fresh rosemary. Combine with a 750 ml bottle of Owney's rum in a large Mason jar or simply add the sprig directly into the bottle. Let sit in a cool, dark place for approximately two weeks, give or take a few days, depending on your taste.

Bridget Firtle did an experimental little run of a few flavored rums for the holidays in 2013—a project that she plans to repeat in coming years. Of the 2013 offerings, her rosemary-infusion was our favorite, and she was kind enough to share a home version of the recipe with us. The cocktail we designed for it is loosely based on a drink called the Riposto, which was featured in *The New Brooklyn Cookbook* by Melissa and Brendan Vaughn. The drink originated at Brooklyn Social, a sister establishment of Henry Public, as a vodka drink. But we really like the piney funk that the rosemary-infused rum brings to the table. It's a fun, festive drink for your next holiday party.

BED-STUY BRIBE

1 oz Owney's rum
1/2 oz Zaya 12 year old Trinidadian rum
1/2 oz Cappelletti Aperitivo
1/2 oz lime juice
1/2 oz simple syrup (*see page 11*)

Place all ingredients in a cocktail shaker mostly filled with ice. Shake hard and strain over fresh ice in a collins glass.

This cocktail is an original of Guillermo Bravo's from The Richardson. Cappelletti is a bitter wine-based aperitivo that's kind of like a cross between a vermouth and Campari. When it debuted in the U.S. market in 2013, Guillermo wanted to try to pair it with the Zaya rum. "I wanted a summer rum drink that had a sweet rum body but bitter finish to it, something I've wanted Daiquiris to have.

"The name came from where all the names for my cocktails come from: music. I had been listening to Rancid's *Life Won't Wait* album. The lyrics reminded me of stories that could very well take place in Bed-Stuy, where I live. Through a bit of word association, the name 'New Dress' (one of the songs on the album) became Bed-Stuy Dress, then Bed-Stuy Bride, and finally Bed-Stuy Bribe. The drink ended up on our winter menu, and Bed-Stuy Bribe has a colder edge to it and represents Bed-Stuy in the winter more effectively."

100 YEARS

AMERICAN OAK

NEW YORK DISTILLING COMPANY

DSP NY - 15027

BARREL #028

Char #3

Picture this: an educated entrepreneur, fed up with the world of business, wanted to create an exciting, alcohol-related venture in an up-by-your-bootstraps manner, ahead of the curve, and in the all-embracing borough of Brooklyn. Are we talking about one of the distilleries that emerged in the late 2000s? No. We're talking about Brooklyn Brewery, founded by Tom Potter in 1988, back when several of today's local craft distillers were still wearing short pants! ▶

Potter wanted to make craft beer and went at it full throttle. He opened a brewery in Williamsburg, once the American epicenter of beer brewing. Master mixologist DeGroff says that at the time, the idea of a brewery in Brooklyn wasn't exactly a no-brainer. "A friend of mine was asked to invest in Brooklyn Brewery, and he said, 'Get out of here, a brewery in Brooklyn? That's a terrible investment. It'll never work!' He's kicked himself in the ass ever since!"

Years later, after retiring from Brooklyn Brewery, Potter had an idea for a new venture—he sensed that the micro-distillery scene was in a very similar place to where microbreweries were 25 years earlier. Also, he wanted very much to start a business from scratch with a new partner: his son, Bill. So he returned to the borough he knew best, Brooklyn, and began researching the distilling business.

Meanwhile, Allen Katz was intimately involved with some of the most important, and certainly coolest, trends in the contemporary culinary world. Over the previous ten years, he had been chairman of Slow Food U.S.A. and a board member of both the New Orleans Culinary & Culture Preservation Society and The Manhattan Cocktail Classic, and he even hosted "The Cocktail Hour," a weekly program on Martha Stewart's SiriusXM Satellite Radio. But it was a trip to England that first got him thinking of opening a distillery. Said Katz recently, "For me, the fantasy of opening this distillery goes back 10 years. I was on a trip to Plymouth, England, and I went to the Plymouth Gin distillery—I had never heard of navy-strength gin." It was eye-opening to Katz that such important styles of spirits from distilling's past had become so rare and hard to find.

So he set to planning and becoming more involved in the spirits business, seeking advice from friends—one of whom happened to be a friend of Potter's, too.

"He said, 'You guys are talking about similar ideas. You should get together.'" Over lunch, Potter and Katz explained what their individual roles would be and agreed that working together made sense. Katz said, "Tom's acumen for developing small businesses was crucial. And my expertise in the spirits and cocktail world made us a great match."

Potter recently told Robert Simonson in a piece for EdibleBrooklyn.com, "I give Allen tons of credit for having ideas from the start for coming up with gins that were unique but still fit within that classic profile." Katz and the Potters founded New York Distilling Company and chose a space for their new factory in the

"We want to be purposefully different. We want to make spirits that are relevant, that taste really good, and that add to the conversation of categories that already exist."

—ALLEN KATZ

Brooklyn neighborhood Potter knew so well, Williamsburg.

The three of them together are not only making great spirits, but they are crafting expert odes to New York's past. "Every business model is different," says Katz. "People's personalities are different. We want to be purposefully different. We want to make spirits that are relevant, that taste really good, and that add to the conversation of categories that already exist."

For New York Distilling Company, that meant reaching back into history to revive some older categories—spirits that were commonplace in 19th-century Brooklyn saloons but had all but vanished from the scene, like Dutch-style gin, rock and rye, and of course navy-strength gin. Explained Katz, "It's the strongest proof gin in the world, made at the strength demanded by the British Royal Navy. The idea was that even

if the gin spilled on gunpowder, the powder would still ignite. Navy strength gin has an interesting history; it was wildly successful in New York but hadn't been widely sold here in nearly a century. Nobody really knew about it, but it had this great New York connection, and it's a great spirit that people from all over can appreciate. That's what we're looking for: products and brands that have a bandwidth that strives for local relevance, but survives well past that."

Of all the local spirit-makers, New York Distilling Company has made the most inroads at the local bars and restaurants. There was an embarrassment of riches to choose from when putting together the cocktails for this chapter. Their products are great, they connect to a meaningful narrative, and they have the professional backgrounds and expertise to keep expanding and improving.

THE SHANTY

(Williamsburg)

Not satisfied with just making elegant, classic spirits, the gang at New York Distilling Company is also pouring them in a craft cocktail bar next door to their distillery, called The Shanty. Katz is quick to highlight the contributions of another member of Team NYD. "The credit for our cocktails goes entirely to our head bartender, Nate Dumas. He's one of the great bartenders in the city. He also went to brewing and distilling school, so he has a unique perspective. Nate was working at Clover Club with Julie Reiner. Julie knew what we were trying to do here and said Nate might be an interesting guy to work with."

Dumas explained how his background helps him contribute at New York Distilling. "The people who come into The Shanty often want to know how the gins are made, how the rye is made, and I can answer their questions in detail."

Katz said of the idea for The Shanty, "In our initial discussions I don't believe there was any talk of opening a bar. It was in our own discovery process that we learned it was a possibility. We thought if we could find the right space, a bar would be something interesting to pursue."

But according to Katz, The Shanty isn't meant to just be a distillery's pub. "We have 200 bottles behind the bar; four of them are ours. We are open seven days a week as a neighborhood bar. As a start-up, it's revenue for us daily, but it's more than that. We're cocktail people making spirits that are best utilized in cocktails, and The Shanty gives us an opportunity to demonstrate that skill and say to the public, 'Hey, here are some things you can do with our stuff.'"

THE 700 SONGS GIMLET

1 1/2 oz Perry's Tot Navy Strength Gin
3/4 oz fresh lime juice
1/2 oz simple syrup (*see page 11*)
1/4 oz Cinnabark Syrup (*see page 171*)
4 dashes Bittermen's Hellfire Shrub

Shake ingredients over ice and fine-strain into a chilled cocktail glass.

Speaking of classics, there's the Gimlet. For the gin, the choice is Perry's Tot. Owner Allen Katz explains why: "'Perry's Tot has a wonderful structure for a Gimlet because at its foundation it is based on classic gin botanicals. At 114 proof, it has the strength to mix well in a classic sour so that you can still taste the gin even when it has been appropriately diluted." In reinventing the Gimlet, The Shanty is pairing it with cinnabark syrup and a spicy shrub. The lime is still in there, but this is not your father's Gimlet. It's a spicy, assertive, spirit-forward reinterpretation of a "go-to" cocktail.

Creator Nate Dumas shared his view on why the flavors work. "I think this drink is an example of the importance of carefully selecting the right gin for the right occasion, building on that, and then giving it a little bit of a surprising tweak at the end. The Perry's Tot is rich, weighty, and earthy, and it provides a solid base. Lime, cinnamon bark syrup... nothing out of left field at all. What I think sets it off is the addition of the Bittermen's Hellfire Shrub, which is made with habanero peppers. I love it for its intriguing fruity note even more than I do for the subtle heat that it brings to the table. It's the little something that you just can't quite identify."

The local wildflower honey in Perry's Tot gin does a wonderful job of tempering the aggressive shrub in this drink and rendering it, as Nate says, "subtle." The spices in the botanicals of the gin (cinnamon, coriander, cardamom) are highlighted by the cinnabark syrup, creating quite a bit of complexity and texture.

A GIMLET BY ANY OTHER NAME

With all due respect to the 700 Songs Gimlet, a delicious drink in its own right, it's not technically a gimlet. A gimlet isn't just made with sugar and fresh lime juice; it's made with lime cordial. On the downside, Rose's Lime Juice, the world's most popular commercial lime cordial, isn't something you want in your drinks (it's laden with high-fructose corn syrup and other artificiality). Fortunately, our friend Toby Cecchini of Long Island Bar, a renowned spirits expert and historian, has worked up an excellent version you can make at home. For the record, to make a classic gimlet, simply combine 2 ounces of gin and 3/4 ounces of homemade lime cordial. Shake and strain into a cocktail glass. Or you can make a Raymond Chandler gimlet, which is 50/50 gin and lime cordial. In *The Long Goodbye*, Philip Marlowe claims, "It beats Martinis hollow."

LIME CORDIAL RECIPE

18 ripe limes, preferably organic
2 1/2 cups sugar

Recipe adapted from Toby Cecchini. Wash and dry limes, then remove their zest carefully, keeping as little white pith as possible. Slice the limes in half and juice them. You should end up with approximately 2 1/2 cups of juice.

In a re-sealable container, add the sugar to the juice and stir until combined. Squeeze the peels over the juice and drop them in. Stir to combine. Cover and allow to macerate for 24 hours. Strain through a fine strainer and decant into a clean, capped bottle.

WEREWOLF BAR MITZVAH

1 oz Perry's Tot Navy Strength Gin
1 oz Pierre Ferrand 1840 Cognac
1/2 oz Denizen Rum
1 oz fresh pineapple juice
1 oz fresh lemon juice
1 oz fresh orange juice
1/4 oz simple syrup (*see page 11*)

Shake with crushed ice and pour unstrained into a chimney glass. Float 1/4 inch of Manischewitz over the back of a teaspooon on top of the drink and garnish with an orange wedge, a mint sprig, and a cocktail umbrella.

Werewolf bar mitzvah spooky, scary; Boys becoming men, men becoming wolves

From the moment the six-second sketch, "Werewolf Bar Mitzvah" aired on the NBC sitcom *30 Rock* in 2007, it inspired spoofs, t-shirts, and naturally a Tiki drink from Fort Defiance's Zac Overman.

In the sketch, the character Tracy, played by Tracy Morgan, flashes back to the video of a novelty song he did in 1986 that was meant to be an anthem for the bar mitzvah circuit. Overman said, "I knew with the name I wanted to use Manischewitz somehow, which is a difficult thing to do, but the cream sherry float in Trader Vic's Samoan Fog Cutter seemed like a good place to start." Overman started with the Samoan, then led it through a rite of passage to give us the Werewolf Bar Mitzvah. Oy Vey.

ANOTHER PRETENTIOUS BROOKLYN COCKTAIL

2 oz Dorothy Parker American Gin
3/4 oz Kina L'Avion D'Or
grapefruit twist

In a mixing glass or shaker mostly filled with ice, stir the Dorothy Parker and Kina L'Avion D'Or. Express a grapefruit twist into a rocks glass and strain the cocktail into the glass.

This meta-cocktail is the brainchild of our friend Alicia Blegen from Marietta in Clinton Hill. Think of the Kina L'Avion D'Or—a quinine liqueur—in this cocktail as an elevated (and inebriated) tonic water, which when mixed with Dorothy Parker Gin creates a gin lovers delight. This soft, floral gin features juniper front and center on the nose, balanced—but not overpowered by—the sweet scent of hibiscus petals and citrus oil. The subtle spiciness of cinnamon and cardamom in the body of the spirit breaks the bitterness of the cocktail just enough to allow the lingering of fresh citrus notes through the finish. The end result is mash-up of two gin classics: the Gin and Tonic and the Martini.

FORT DEFIANCE
TIKI NIGHT

(Red Hook)

At first, Zac Overman was just a regular at Fort Defiance—even though he had to take the bus to get there. That should give you an idea of what kind of loyalty the place inspires. In his own words, Zac "wore down" owner St. John Frizell to the point that he was allowed to step behind the bar and learn the craft. It must have worked out OK, because these days Zac is the bar manager. We're big fans of his Tiki night. "My wife and I eventually moved to Red Hook, and about once a year, we'd just decide to throw a Tiki Party. We made mid-century snacks and I put together a menu of eight drinks with a little history on the menu, and we'd invite all our friends over, dance, drink too much, and have a really great time."

Eventually St. John appealed to Zac to do a Tiki night at Fort Defiance to spruce up a random Thursday night and get people in the door.

Zac was thrilled. "In my head I thought 'We're going to do this every week.'"

"I told Zac I wanted to do one once a month," St. John explained. "Zac said, 'OK,' but then he handed me next week's menu."

Speaking for ourselves, we are happy for the not-so-hostile takeover. The Tiki-inspired food is delicious. The menus are beautifully designed and fun. The music—carefully curated by Zac—is spot on. And best of all, as Zac says, "We have all these ridiculous mugs and I wear a dumb hat."

> # "This is a cocktail that celebrates New York— past, present, and future."
> ## —ALLEN KATZ

THE CHIEF GOWANUS MARTINEZ

1 1/2 oz Chief Gowanus New-Netherland Gin
1 1/2 oz Carpano Antica Formula sweet vermouth
1 tsp maraschino liqueur
2 dashes Angostura aromatic bitters

Place all the ingredients in a cocktail glass or pitcher mostly filled with ice. Stir and strain into a coupe or cocktail glass. Garnish with a lemon twist.

To drink a Martinez made with Chief Gowanus New-Netherland Gin is to experience 200 years of New York history in a glass. The Martinez is an early relative of the Martini and the Manhattan that initially featured sweet (or "Italian") vermouth and the sweeter style of gin popular in the 19th century. Until recently, it was nearly impossible to make an authentic recreation of the Martinez, because it was hard to find a sweeter Dutch-style gin in America. In his classic book *The Craft of the Cocktail*—certainly one of the definitive works on the art of bartending—Dale DeGroff even replaced the "Old Tom Gin" of the original Martinez recipe with "gin with syrup" out of the need for something true to the originally intended flavor. Zac Overman of Fort Defiance, who made this drink for us, explained to us why the combination works so well: "This drink is just a natural fit for the Gowanus, which has more backbone and spice than a typical dry gin— it stands up really well to the Carpano."

Aged for three months in seasoned barrels, Chief Gowanus is imparted with vanilla flavor from the oak, whose sweetness perfectly rounds the spicy and bitter qualities of the rye and juniper. When married with a full-bodied vermouth, it's a cocktail flavor bomb exploding with hints of juniper, dried tangerine peels, and a light whiskey spice, ending with a lingering hops flavor.

FIRST NIGHT BACK IN LONDON

1 1/2 ounces Dorothy Parker American Gin

1/2 oz Coruba Jamaican Rum

3/4 oz Asian pear/red wine reduction

1/2 oz lemon juice

1/4 ounce simple syrup (*preferably made with Demerara sugar, see page 11*)

2-3 cilantro leaves

dash Angostura aromatic bitters

Add all the ingredients to a cocktail shaker mostly filled with ice. Shake vigorously, and double-strain into an ungarnished coupe.

ASIAN PEAR REDUCTION

Chop up an Asian pear, discard the core, and heat pear chunks in 1 1/2 cups dry red wine. Cook on medium heat until the wine reduces to around a cup. Strain and cool.

We like any cocktail that gets its name from a Clash song. And it's appropriate that this one comes from Dram, one of the great cocktail bars of Brooklyn, where on a typical night you can watch a subtitled B movie on a 6-foot-by-6-foot screen and hear all manner of punk rock on the stereo. Despite the audio-visual stimulation, it's a great, relaxed atmosphere, with an open kitchen station busting out terrific bar bites to go with your drinks.

This unusual cocktail, created by Jeff Hazell of Dram, is a winner. We wouldn't have thought to pair the Dorothy Parker with such bold, complex, tannic flavors, but the combination works.

Matthew Bellinger, a bartender at Dram, describes the Dorothy Parker perfectly. "The floral quality of the Dorothy Parker really helps brighten the whole drink. A lot of new American gins really dial back the juniper and are super floral. Others are really barreling down on the juniper, trying to get back in touch what they feel a London dry gin should taste like. The Dorothy Parker has a balance of both ideas and a really unique flavor as well."

CANARSIE WARRIOR

1 oz Chief Gowanus New-Netherland Gin

1 oz Perry's Tot Navy Strength Gin

1/4 oz Amaro CioCiaro

1 tsp Orgeat Works Toasted Orgeat Syrup

dash Angostura aromatic bitters

Stir with ice and strain over fresh ice in a frozen rocks glass. Garnish with a lemon and orange.

This drink, designed by The Clover Club's Travis St. Germain, is basically a gussied up gin Old-Fashioned. Says Travis, "Because of the rye base of the Chief Gowanus, the Canarsie Warrior has a more familiar tone than a lot of other Old-Fashioned-style drinks that use a junipered neutral grain spirit." It's a drink that really sings in two-part harmony, with the rye in the Gowanus and the Angostura aromatic bitters keeping the drink grounded in its origins, and the Perry's Tot amping up the appeal to the gin drinker. But what about those wonderful background notes? Travis explained to us, "As for the flavors working so well together, I've always been a fan of the high orange notes of CioCiaro alongside Adam Kolesar's lightly toasted orgeat. Once that was settled, it was pretty easy from there to find the proper balance."

CLOVER CLUB

(Carroll Gardens)

Clover Club owner Julie Reiner has become a big supporter of the local spirits scene. "We have a section on our menu now, called Hometown Hooch, which is dedicated not just to Brooklyn, but to all of New York State."

Reiner hasn't become one of the most successful bar craft cocktail owners in New York by accident. She understands the things that really matter, like creating a great atmosphere, providing great service, and using the best quality ingredients to make drinks. So being local isn't enough to get on her menu. "All of a sudden things that are made in Brooklyn are the hippest thing ever right now. And I'm glad that I live in and have a business in the hippest borough—but just because it's made in Brooklyn doesn't mean it's good. But we're so happy to have these locally made spirits that we're excited about. I'm a huge supporter of Allen Katz and the stuff he's doing over at New York Distilling Company, and we're big fans of Owney's rum as well. There are plenty of other interesting local products out there as well, like Adam Kolesar's Orgeat Works line."

THE DEVIL'S DUE

2 oz New York Distilling American Rye
1 oz aged premium rum
1/2 oz Cynar
orange twist (*expressed over the drink
and then dropped*)

*Build over rocks in a rocks glass. Give it a good
stir. Flame an orange peel by holding the edges
of a thick twist and squeezing it through a
flame, as demonstrated in the photos.*

(*Amounts will vary with the size of the glass.*)

Dale DeGroff helped revitalize the state of the American cocktail while working at The Rainbow Room, and he influenced a generation of bartenders with his must-own book, *The Craft of the Cocktail*.

For this cocktail Dale chose NYD American Rye as the main spirit. This three-year-old straight rye whiskey has a mild nose and a taste of rye that veers pleasantly towards the sweet side, with the full character of rye grain displayed. The typical spice notes found in rye whiskey are blended well with the sweetness of malted barley and corn in the mash bill.

Dale explained how this cocktail came together: "The rye is pretty rich, but also pretty bold. The Zacapa 15 has that character that I want and it essentially finished the whiskey in the glass. You could use Zacapa 15, or Appleton 12 Year Old, or Extra Old Mount Gay—any nice aged premium rum will do the same job."

THE HAPPY ROCK

1 1/2 oz Mister Katz's Rock & Rye
1/4 oz green Chartreuse
1/2 oz Pierre Ferrand Ambre Cognac
1/4 oz Mathilde Pear Liqueur

Shake ingredients over ice and strain into
a rocks glass filled with one large ice cube.
Garnish with a broad orange twist.

Henry Miller, known as the "Buddha of Brooklyn," spent his formative years in Williamsburg, which he later chronicled in the classic of American literature, *Tropic of Capricorn*. While living in Paris in the 1930s, Miller was dubbed by his friends the 'Happy Rock,' a reference to his "sublime indifference," a yearning for a life with no hopes and no regrets. It was during this time in Paris that Miller developed his signature semi-autobiographical style, which blended philosophy, mysticism, and explicit sexual content; it heavily influenced generations of writers and horny teenagers alike. His tour de force *Tropic of Cancer* was first published abroad in 1934 but wasn't allowed to be published in the U.S. until 1961. In the end, Miller saw his censorship as validation. He said, "The book is living proof that censorship defeats itself."

According to Miller's daughter Val, late in life the Happy Rock drank a glass of the French liqueur green Chartreuse every night, a habit he brought with him from France. It's fitting that a Williamsburg distillery would lend its spirit to a cocktail in honor of the neighborhood's most famous son.

THE HOTEL BOSSERT'S BROOKLYN COCKTAIL

2 oz Dorothy Parker Gin
1 oz sweet vermouth
dash Hella Bitter Aromatic bitters

Place all ingredients in a cocktail
shaker mostly filled with ice. Stir well,
strain into a cocktail glass, and twist a
lemon peel over the top.

This one comes courtesy of the great David Wondrich himself, who wrote a wonderful essay on the history/evolution of the Brooklyn cocktail for *Edible Brooklyn: The Cookbook* by Rachel Wharton.

There have been many versions of the Brooklyn along the way, many of them suspect. But there have also been several interesting variations—see pages 32 and 228 for two other examples.

Wondrich says of this one, "In the 1930s and 1940s, the roof garden of the Hotel Bossert on Montague Street in Brooklyn Heights was one of the most popular night spots in the city. This is the Brooklyn Cocktail invented there in 1935 when the bartenders got sick of mixing up Manhattans for visiting Manhattanites."

DEATH BED MANHATTAN

2 oz New York Distilling American Rye
1/2 oz Punt e Mes sweet vermouth
1/2 oz Carpano Antica Formula sweet vermouth
2 dashes Angostura aromatic bitters

*Stir ingredients over ice until
exceedingly well-chilled. Strain into a
chilled cocktail glass and garnish with
a brandied cherry on a cocktail pick.*

"The classics are classics for a reason," Julie Reiner told us on one of our trips to Clover Club. "They're simple, easy to make, and they're great." Allen Katz tends to agree. When we asked him what his favorite cocktail was, he answered without hesitation, "Manhattan. I'm a big rye whisky fan. I like the taste. I like the history. I like the sense of authenticity as an American whiskey. You can ask my wife: it's my death bed cocktail."

The young guys from Industry City Distilling have been described by one wag in the Brooklyn cocktail crowd as the "boy band" on the local scene. To find the truth in that statement, you have to reach back way beyond the likes of One Direction or New Kids on the Block to find the proper analogue. Maybe if you consider the mid-60s Rolling Stones or Beatles a "boy band" you might be getting closer to the mark—because like those revolutionary groups, the ICD collective is taking an idea that's old and completely reinventing it for the modern world. ▶

One of the great benefits of the DIY approach is that you can do it any damn way you please. That's Industry City Distilling in a nutshell. This group of young men started an industrial cooperative called City Foundry for the purpose of "improving small-scale manufacturing processes through the blending of science and art." Then, while embarking on a project to create carbon dioxide for aquatic ecosystems (fish tanks), they recognized that they had a good model for distilling alcohol. So they decided to reinvent the distilling process from scratch with the stated goals of maximizing efficiency and minimizing waste. They rented space in Brooklyn's historic Bush Terminal and set to work creating a process of distilling beet sugar into vodka unlike anyone had ever tried before.

When we arrived at the distillery, Peter Simon insisted that we should tour their unique

production facility. Before joining the team at Industry City, Simon was a yoga instructor who earned extra money waiting tables. One day he served a cup of coffee to the man who would become the company's co-founder, Dave Kyrejko. Simon went on to hold multiple jobs at ICD, including head of outreach and also chief blender.

When the tour brought us to the guts of the distillery—the bioreactors—Kyrejko took over. After all, he's the distiller, engineer, yeast propagator, and designer of the equipment. In short, Industry City Distilling is his baby, and a proud papa he is. Kyrejko picks up the Process 101 talk on fermentation and distillation and boils it up to a PhD discourse more likely to be presented at a TED conference than in a loft overlooking Gowanus Bay. There's nothing easy about what he's explaining, but his approach is as clean as his vodka. Kyrejko

elucidates complicated subjects with the ease and appeal of Neil deGrasse Tyson on *Cosmos*.

The back leg of the tour circled through the machine shop where a third co-founder, Zachary Bruner, designs and creates the tools and equipment they use to make their vodka. Where a craft distillery typically spends upwards of $500,000 on its distilling equipment, because of this workshop, Bruner was able to build what they needed for $600.

Bruner is quiet and industrious, but when a question came up about the shop, he was quick to answer with thoughtful remarks. These tools are extensions of him and his imagination. As he put it, "It's a dream to be able to make these machines from scratch." Bruner met Kyrejko at a Rhode Island School for Design summer camp. Somehow the two knew that they had

complementary qualities: Kyrejko designs it and Bruner builds it. On a shop table we noticed a glass gallon bottle with a taped label that read "solvent." It was the part of the still run that Simon didn't like for drinking, but it made a great cleaner for Bruner.

When we finished the tour, we met the last member of the Fabricating Four, Max Hames. Hames was finishing a meal made in the workshop kitchen with a stove that burns more alcohol that Simon didn't like. How Hames came to the group is a story straight out of a sitcom plot. He came to visit the distillery one day thinking it was already built. Finding an empty loft and willing to help, he stuck around to help the boys build it and eventually was put in charge of operations management. He just barely had time to finish a bowl of food and pose for a group picture for us before he had to get back to work.

THE CHALLENGES OF VODKA

"The fish tanks were absolutely gorgeous, but nothing really brings people together quite like alcohol."
—DAVE KYREJKO

Kyrejko talked about the difficulty, scientifically speaking, of what goes on at Industry City. "Making real vodka is really hard. If your goal is to make a vodka without any charcoal filtrations, single pass under distillation, that's a hell of a challenge." And that's not to mention the hurdles on the marketing/sales side. The truth is that as popular as vodka is with the general public, in cocktail circles it's a dirty word.

Part of the problem, Kyrejko believes, stems from the quality of vodka that's currently available: "The majority of the vodkas on the market are all from the same source, an industrial ethanol that you buy and mix with water. They're not distilleries at all, they're rectifiers, which means they don't actually make their product other than adding water to the alcohol. And they're not good. People tend to say, 'Oh, vodka reminds me of rubbing alcohol,' and they're absolutely right. The industrial spirits can have up to three percent isopropanol in them, which is rubbing alcohol. Humans have an amazing nose, and they recognize it right away."

But what about the idea that vodka is supposed to be tasteless and odorless? "That's just a legal definition that doesn't really mean anything," Kyrejko explained. "If you pay attention to our vodka, there is plenty of flavor."

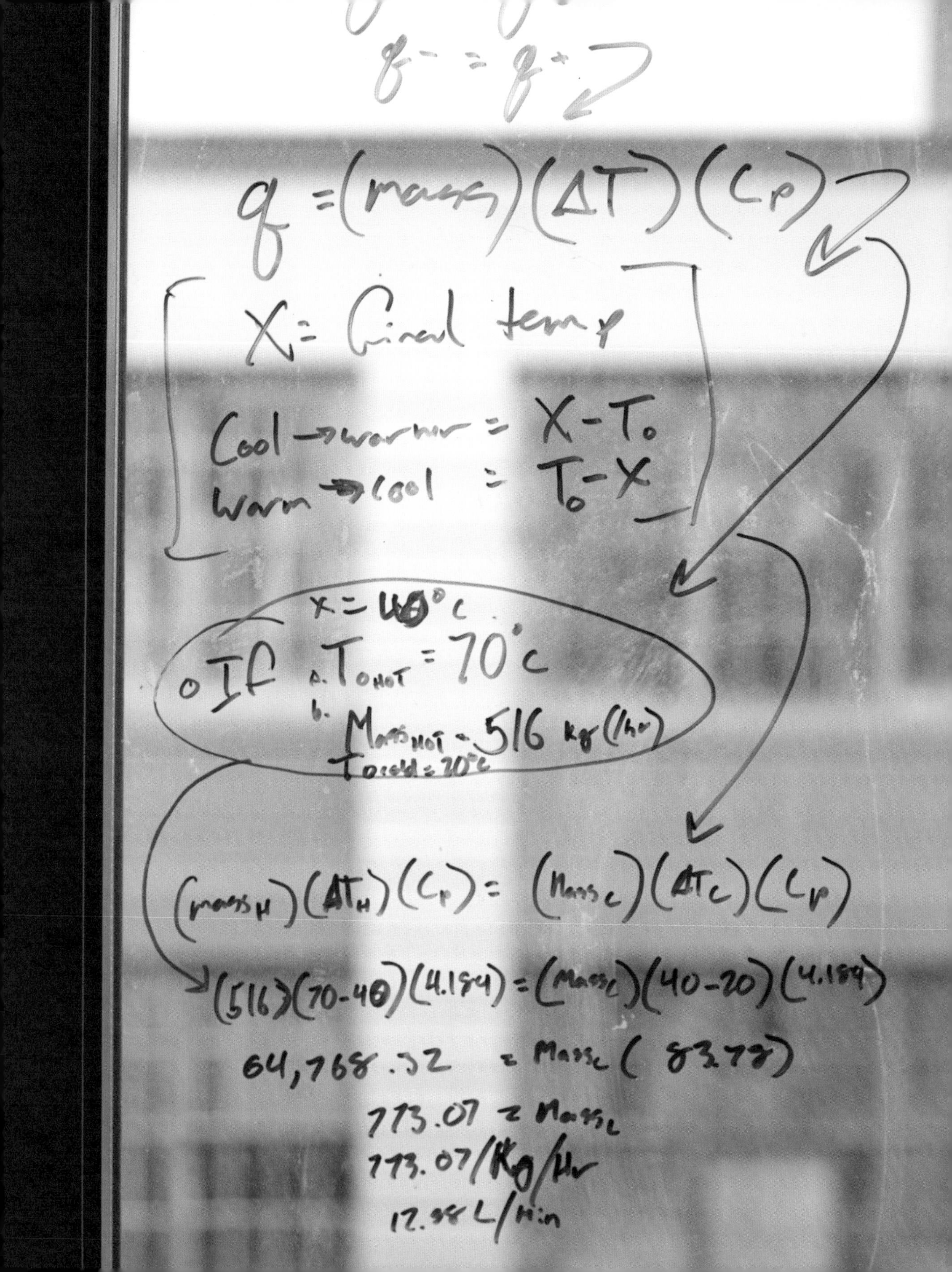

$$q^- = q^+$$

$$q = (mass)(\Delta T)(C_p)$$

$X = $ final temp

Cool $\rightarrow$ warmer $= X - T_0$
warm $\rightarrow$ cool $= T_0 - X$

o IF
$X = 40°C$
a. $T_{0\ HOT} = 70°C$
b. Mass$_{HOT} = 516$ kg (/hr)
$T_{0\ cold} = 20°C$

$$(mass_H)(\Delta T_H)(C_p) = (mass_C)(\Delta T_C)(C_p)$$

$$(516)(70-40)(4.184) = (mass_C)(40-20)(4.184)$$

$$64,768.32 = Mass_C (83.78)$$

$$773.07 = Mass_C$$

$$773.07 / Kg/hr$$

$$12.98 L/min$$

"I don't know if you've been to the other distilleries yet, but we don't really have much in common," said Kyrejko. That's for sure. At ICD, we didn't see any of what's considered standard distillery equipment: copper stills, forklifts, giant fermenters. "We make our equipment in-house because what we need simply doesn't exist. We had to make pretty much everything."

Once they had their equipment in place, it was time to tackle the raw materials. Kyrejko said, "I pretty much shut myself away in the lab and worked on a new piece of technology called a mobilized cell bioreactor. What that does is it lets you take all the power of those giant fermentation vessels and put it into a very small container."

What Kyrejko was tackling was at the heart of everything at ICD: efficiency. Kyrejko and company wanted less wasted energy, less wasted organic material, just less waste.

We had seen photographs of these famed reactors on the website and elsewhere, but the photos don't do them justice. One of the most impressive things about the system is that it is incredibly compact—and scalable. It could easily be bigger than it is, and it looks as though it will be at some point.

Kyrejko continued, "So these things can take 85 gallons of high sugar content, 14 percent alcohol potential wash, and turn it into 14 percent alcohol, every 24 hours." According to Kyrejko, a standard distillery set-up would take two weeks to perform the same task.

Another special aspect of Industry is the way they make and propagate their yeast—it's all done in-house, and they've devised a way to encapsulate the yeast into a bead-like amalgam, which allows it to live actively for weeks on end! This allows ICD to make alcohol 24 hours a day, for months at a time.

Next, the good alcohol is separated from everything else, and that's where Ivy and Betty come into play. Those are the nicknames of the two stills that convert the fermented wash into vodka. ICD uses a unique two-stage distillation process that doesn't resemble anything else we'd seen or even heard of. Kyrejko explained, "Your typical copper pot still is woefully energy-inefficient. We use copper, but we use the copper where it's important, so you'll see that we have copper in our coils, and what copper does, is it actually absorbs sulfur-bearing compounds and makes a much cleaner spirit. So there is a science behind your copper pot stills, but that's where it ends; you only need a tiny amount of copper."

Kyrejko told us that in his view, the chief problem with a pot still is that it takes hours

to process, and during that time you've got all these different compounds being boiled over and over again. "So what this lets us do, is to cut the distillation time from hours to seconds, and what we do is is we get rid of the boiler."

Instead of the boiler, the process uses a separation column (the aforementioned Ivy). Steam is fed through the bottom of the column while the wash is pumped into the top. "You've got wash coming down, steam going up, and inside there are these little pieces of metal called 'packing.' And as the steam goes up, it makes the packing really hot, and as this wash hits the packing, the alcohol flashes off, goes upward, and the water goes downward." That's how the alcohol is separated off from the rest. In traditional distilling, potentially hazardous "heads" (the early part of the run) and "tails" (the late part of the run) are separated from "the

hearts" (the good stuff). But ICD takes this to another level, taking 40 separate cuts once every 20 minutes, marked by a bell. Simon tastes through the various cuts, separating the good from the bad no matter where in the run it occurs.

The real "art" part of the process comes in the blending step. Simon explained to us what he does to sift through each cut to determine exactly which will go into their vodka. "You take those 40 cuts, you bring them in here, and you taste through them all; you mark them down for taste, texture, smell, and so you get this really good blend of science and art."

Surprisingly, to us anyway, there are often cuts in the middle of the run that Kyrejko describes as "completely not cool" that get removed every time.

It's a triumph of both product and process. Kyrejko summed up the approach succinctly: "By getting rid of the whole craft mindset and looking more towards industry, we brought a whole new system to the game."

While it is impossible to reinvent the distilling process—yeast eats sugar and makes alcohol—it is possible to reimagine and improve it. The goal behind Industry Standard Vodka is a pure, clean expression of alcohol. The cocktails in this chapter—like so many throughout the book—have been designed around the premise that the base spirit should take center stage in anything you drink. What's more, in the case of Industry Standard Vodka, the base spirit should be elegant and refined, an expression of the equipment in the workshop that creates this ethereal liquor.

COSMOPOLIS

While most bartenders in this book are updating classic cocktails through classic techniques, there is an avant-garde in the cocktail world that is reworking the recipes through modern food science. The Cosmopolis is a new view of the classic Cosmopolitan with Industry Standard Vodka in mind. In this version, the cranberry juice, lime cordial, and triple sec are formed into delicate spheres using a technique borrowed from molecular gastronomy. (Instructions for making spheres are readily available online.) They float about the vodka, barely influencing its flavor, hinting at the ultimate symbol of 90s cocktail culture, but updating it to suit a vision of the future. It's a taste of tomorrow in a glass of yesteryear.

THE BROOKLYN VESPER

2 oz Industry Standard Vodka
1 oz Dorothy Parker Gin
1/2 oz Kina L'Avion D'Or

Put all ingredients in a mixing glass, STIR (apologies to Mr. Fleming), and strain into a 60s-style cocktail glass— unless you happen to have a deep Champagne goblet lying around.

"A dry Martini," [Bond] said. "One. In a deep Champagne goblet."

"Oui, monsieur."

"Just a moment. Three measures of Gordon's, one of vodka, half a measure of Kina Lillet. Shake it very well until it's ice-cold, then add a large thin slice of lemon peel. Got it?"

"Certainly, monsieur." The barman seemed pleased with the idea.

"Gosh, that's certainly a drink," said Leiter.

Bond laughed. "When I'm...er...concentrating," he explained, "I never have more than one drink before dinner. But I do like that one to be large and very strong and very cold and very well-made. I hate small portions of anything, particularly when they taste bad. This drink's my own invention. I'm going to patent it when I can think of a good name."

That dialogue is straight from Ian Fleming's novel, *Casino Royale*, describing the drink that came to be known later in the novel as the Vesper, after the book's heroine, Vesper Lynd.

Kina L'Avion D'Or is an aperitif made by the hooch historians at Tempus Fugit Spirits. Initially conceived as an effective delivery system for quinine, the original Kina Lillet (before the recipe was changed in the 1980s) had a common flavor profile to Kina L'Avion D'Or. That makes it a great replacement for Kina Lillet in the Vesper. To make a Brooklyn Vesper, we've altered the proportions a bit from the Bond/Fleming version.

Thanks to Henry Lopez at Lucey's Lounge for giving us the idea for this one, though he makes the drink differently than we do. For one thing, he listens to Fleming and shakes. Like us, he flips the amounts on the vodka and gin, but he uses Cocchi Americano, not Kina L'Avion D'Or. Either works, but we prefer the Kina (not least because we just made you buy a bottle to make Another Pretentious Brooklyn Cocktail back on page 191.)

LUCEY'S LOUNGE

"I was a red-headed boy named Lucey growing up in Brooklyn and I can tell you with certainty: that which doesn't kill you only makes you stronger."

Darren Lucey was holding court with us and his business partner Henry Lopez on a Friday morning in December of 2013. Eventually the subject turned to Brooklyn and how it's changed—or not changed, "It's funny to me because everybody's saying, 'Well, Brooklyn's hip now.'" I tell them, 'I hate to tell you, Brooklyn's always been hip. You're just catching up now.' There have always been places out here that were doing things better."

Part of the advantage of opening a business off the beaten path, even by Brooklyn standards, is financial—or, more accurately, how the financial side of the business allows the creative side to flourish. "Our rent here in Gowanus is reasonable; there's not a lot around us now. You can still try to find the deals before landlords try to rip you off. That means you get to decide who you are for yourself. You don't have to let somebody else decide for you."

And what has Lucey's Lounge decided to be? Lots of things. To us, it's the ultimate neighborhood bar. The place has a down-to-earth vibe where you'd feel comfortable ordering anything from a shot and a can of beer to a barrel-aged Manhattan or other of-the-moment cocktail.

Despite being new, the place feels like it's been around forever. To the Wall Street Journal, it's "the greenest bar ever built in New York City." Well-integrated reclaimed elements like 1940s carnival glass and an old strip of Brooklyn sidewalk attest to that. And the owners certainly don't lack for ambition: head bartender Henry Lopez told us proudly, "Our goal is nothing less than to be the best bar in the world."

> "All these small producers are forgoing mass production shortcuts that have steadily diluted the modern palate. The more of them I meet, the clearer it is that we share a place at the forefront of a revolution in how we choose to fill each other's cups."
>
> —DANNY KENT

DANNY KENT'S MARTINI

2 oz Industry Standard Vodka
1 oz Danny Kent's Dry Vermouth (*page 225*)
olive for garnish

Place vodka and vermouth in a cocktail shaker mostly filled with ice and stir until very, very cold. Strain into a chilled cocktail glass and garnish with an olive.

When we first met Danny Kent at the 2013 Brooklyn Shaken and Stirred Event, Locanda Vini e Olii's bar manager was a man on a mission. There was a score of cocktail glasses in front of him, and he was deftly stirring his pitcher of Martinis and filling every one. Pete asked for a taste of his homemade dry vermouth and Danny splashed him some without missing a beat, then went right back to stirring and pouring—all with a smile from ear to ear. Clearly this was a man who loved his work! And his homemade vermouth was a revelation. It was spicy and oily, and there was a fresh, bitter herbaceousness that blended perfectly with the sweetness of the base wine.

If you don't think a vodka Martini recipe belongs in a 21st-century cocktail book, do us a favor: try this one and let's talk again.

DANNY KENT'S DRY VERMOUTH

3 cups moscato D'Asti wine	peel of 1 orange
2 cups water	4 stalks rhubarb (sliced lengthwise in long strips)
1 cup 90 proof brandy (preferably un-aged)	1 lb raw cane sugar
6 tsp chamomile flowers	3 cup wide mouth bottle or Mason jar
6 tsp jasmine flowers	1 cup wide mouth bottle or Mason jar
5 tsp gentian root	funnel
5 tsp cinnamon	coffee filters
4 tsp galangal root	a container, glass or plastic, that can
4 tsp coriander	hold at least 1 1/2 qt
2 tsp allspice	A handsome bottle for your final product
1 tsp wormwood leaf	1 1/2 L of material
pinch saffron	A handsome 1L bottle for your final product

In a sterilized 24 oz Mason jar, add chamomile, jasmine, wormwood, saffron, orange, and rhubarb to moscato D'Asti. This is the base wine of Carpano Formula Antica and is from the same region where the world's best vermouth is made. Let this macerate in a refrigerator for two weeks. *Note:* Make sure to fill the jar entirely to the top before sealing to prevent any oxidation or re-fermenting. Gently shake it once daily to make sure all ingredients are getting a good extraction.

In an 8 oz jar, add brandy, gentian, cinnamon, galangal root, allspice berries, and coriander. This will give you a bitter, aromatic extraction and also serve to fortify—and thus preserve—the wine extraction. Allow the mixture to macerate in a cool, dark, dry place for two weeks, shaking once daily.

After two weeks of maceration, strain and separate all of the solids into a saucepan. (At this point, you may combine the two liquids in a jar. This is unsweetened vermouth.)

To the pan of herb and spice solids, add 1 lb sugar and 2 cups water. Simmer for 20 minutes. Let cool. Strain and separate solids from liquids.

Add half of the syrup to your unsweetened vermouth and taste. Add more to your liking.

When the vermouth is where you want it, strain through a coffee filter twice and funnel into bottles. (*Note:* if you would like your vermouth less cloudy, chill it before straining.)

CLEAR BLOODY MARY

2 oz Industry Standard Vodka

5 lbs tomatoes, quartered

1/2 cucumber, diced

1 celery rib, diced

1 tbsp grated fresh horseradish
(*or 2 tbsp prepared horseradish*)

1 tbsp sugar

1 tsp kosher or sea salt (*more or less
depending on your taste*)

fresh ground black pepper to taste

lemon and lime slices

1 leafy celery stalk for garnish

Purée the tomatoes, cucumber, celery, and horseradish in a blender. Strain through a large sieve or a bag fashioned out of cheesecloth over a large bowl. Get creative here: you're looking for enough surface area so that the tomato mixture can drain freely. This will take awhile. Resist the urge to squeeze the bag or press on the sieve, or you'll end up with red particulate in the drink. It won't taste bad at all, but it will be the wrong color. In the end you should have around one liter of delicious, yellowish tomato water. Add the sugar, then add salt and pepper to taste.

In a tall glass with ice, combine 4 oz tomato water with 2 oz Industry Standard Vodka. Add a squeeze of lemon and a squeeze of lime. Give a stir (or roll into a second glass), and garnish with a leafy celery stalk.

Pete's wife Susan describes this drink as "just like a Spa beverage, but with alcohol." The Clear Bloody Mary—which is really more of a pale yellow, but clear sounds a lot better—is a great, spirit-forward way to feature Industry Standard Vodka. If you go to the trouble of making this for the next brunch you either host or attend, you'll surely win the undying gratitude of your friends. This recipe is adapted from Yvette van Boven's wonderful book *Home Made Summer*. Yvette, please look us up if you ever get to Brooklyn. We owe you one.

A NOTE ON GARNISHES

Normally, we like our Bloody Mary garnish to be a meal. All manner of pickles and olives are welcome. Our friend and food stylist Louise Leonard has even been known to put a quarter of a roast beef sandwich on a pick and stick it right in the drink. But this Bloody is different. Here we recommend the simple approach and lean towards adding just the celery stalk and maybe a slice of raw cucumber and/or lemon/lime slices. The briny flavors that work so well with most bloodies don't do as well with the clean effect of this drink.

JAMES LYONS' BROOKLYN COCKTAIL

GUEST COCKTAIL BY DAVID WONDRICH

In 1945, James J. Lyons, Bronx Borough President, swollen with pride at the global success of the Bronx cocktail, took a swipe at Brooklyn:

"If there is no Brooklyn cocktail, why not conceive one?

Perhaps—
Vinegar as a base.
1 spoonful of raspberries.
1/2 pony of DDT
1/4 branch of a tree
1 ounce of vodka
1 ounce of Durocher bitters.

I do not know how palatable such a Brooklyn cocktail would be, but it would be good enough for the home of 'Dem Bums.'"

I've made a few substitutions and adjusted Lyons's proportions in line with contemporary practice:

LYONS	WONDRICH
Vinegar as a base. *1 spoonful of raspberries.*	spoonful of raspberry-vinegar shrub
1/2 pony of DDT	2 dashes Fernet-Branca
¼ branch of a tree	2 dashes aged branch extract* (*Use any American whiskey that is at least 15 years old.)
1 oz vodka	1 1/2 oz Industry Standard Vodka
1 oz Durocher bitters	2 dashes Bittermen's Xocolatl Bitters
splash of seltzer	splash of seltzer

RASPBERRY-VINEGAR SHRUB
From *The Essential New York Times Cookbook* by Amanda Hesser
Recipe: 1900, Raspberry Vinegar

This recipe originally appeared in the *Times* in an article titled "Women Here and There — Their Frills and Fancies." You may halve or quarter the recipe:

1 cup red wine vinegar
1 1/2 quarts freshly picked raspberries
sugar

1. In a nonreactive bowl, combine the vinegar and raspberries. Cover and let macerate for three days.

2. Mash the raspberries in the bowl, then strain the liquid through a fine-mesh sieve lined with cheesecloth. To every 1 cup of juice, add 1/2 pound of sugar (1 1/4 cups plus 1 tablespoon). Combine the juice and sugar in a saucepan. Bring to a boil and simmer (gently!) for 15 minutes. Let cool, then bottle. Keep refrigerated for up to three months.

3. To serve, add 1 teaspoon raspberry vinegar to a tumbler filled with ice. Add water, sparkling water, rum, brandy or prosecco. Makes about 1 quart.

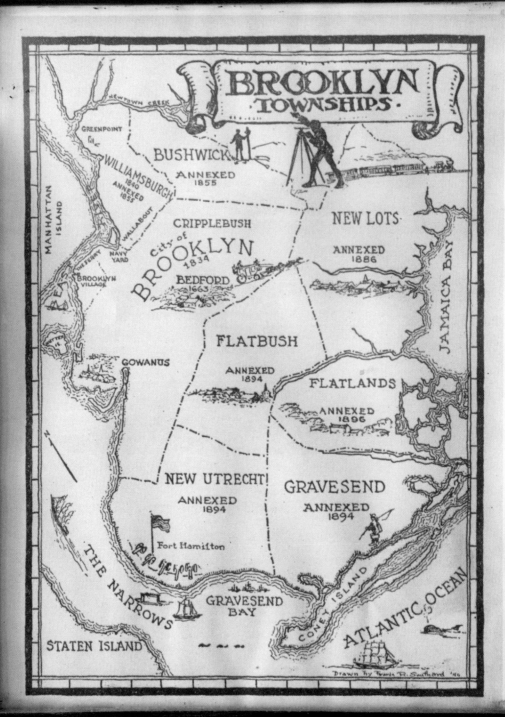

BROOKLYN TOWNSHIPS.

BUSHWICK ANNEXED 1855

GREENPOINT

NEWTOWN CREEK

WILLIAMSBURGH 1840 ANNEXED 1855

MANHATTAN ISLAND

WALLABOUT

NAVY YARD

THE FERRY

BROOKLYN VILLAGE

CRIPPLEBUSH

City of BROOKLYN 1834

BEDFORD 1663

NEW LOTS ANNEXED 1886

JAMAICA BAY

GOWANUS

FLATBUSH ANNEXED 1894

FLATLANDS ANNEXED 1896

NEW UTRECHT ANNEXED 1894

GRAVESEND ANNEXED 1894

Fort Hamilton

THE NARROWS

GRAVESEND BAY

CONEY ISLAND

ATLANTIC OCEAN

STATEN ISLAND

Drawn by Travis P. Southland '40

BACK BEET

2 oz Industry Standard Vodka
1/4 oz Dolin dry vermouth
1/4 oz ounce cold, pickled beet juice
Thinly sliced raw beet for garnish

Add the first two ingredients to a shaker filled mostly with ice. Stir for about 30 seconds until the ice gets all "melty"— that's the scientific term Kyrejko used when he was making this for us. Strain the vodka and vermouth into a chilled coupe or cocktail glass. Measure out the cold, pickled beet juice and pour it on top of the vodka. It should fall to the bottom of the glass. Garnish with a thin slice of raw beet.

BASIC QUICK PICKLED BEETS

1 lb shredded beets
2 cups white vinegar
1/2 cup white granulated sugar
1 1/2 tsp kosher salt
1/2 tsp ground black pepper

Combine vinegar, sugar, and salt in a stainless steel pot and heat to combine. Remove from the heat, add shredded beets, and let stand for 30 minutes. Transfer to Mason jars or other sterilized, airtight containers for storage.

Dave Kyrejko made us his favorite cocktail on a follow-up trip to Industry City Distilling. "It's the simplest cocktail in the world. What I really like about this drink is that the elements of the drink stay separate so you can taste the vodka by itself and then, as you drink it, you get more and more of the beet flavor."

It's a great flavor combination that features the vodka in an aesthetically pleasing and fun way. As Kyrejko says, once the raw beet garnish is applied, "It smells like earth but it tastes like pickles."

Josh Morton owned and operated a computer consulting practice. He and his wife, Susan Weinthaler, lived on Barrow Street in Manhattan where they loved to entertain their friends with fun dinner parties. One day about a dozen years ago, one of Morton's friends introduced him to the world of homemade liqueurs.▶

The friend in question was an American who had worked abroad in Italy as a professional chef. Josh explained, "I was at his house, and he had this stuff in his freezer. One was a limoncello, one was a blood-orange-y liqueur, and there was another liqueur made with rosemary. They were amazing! And he emailed me all the information on how to make limoncello. I started playing around with it and serving it to my friends, and they loved it."

Encouraged by his friends' positive reactions, Morton kept making liqueur. With his dinner party guests as willing test subjects, Morton began to notice a definite pattern emerging. "I started playing around. I did something with blood orange and I did something with lemon. I did something with ginger. Every six months to a year I'd make a batch and it'd be different every time, and I'd make two different batches

with different flavors and the ones with the ginger disappeared first."

This suited Josh just fine. He was growing tired of tearing up his fingers on the Microplane while zesting limes and lemons; grinding ginger was much easier on the hands. Josh's batches of homemade liqueur began trending less toward citrus and more toward ginger, until one day he reached the tipping point: "Out of my own laziness, I would start making batches with less citrus and more ginger... And eventually I made a batch with no citrus, all ginger, and everyone loved it."

The ginger-only liqueur, which became affectionately known as "the ginger stuff" among Morton's circle of friends and dinner guests, also had the novelty factor in its favor; it was unique. But Josh still wasn't thinking of going into business. "I just liked making my friends happy. That was the reason I did this. And for 10 years I made it every six months, every year, and served it after dinner, on the rocks, to my friends. It became a thing. Friends would bring people to a dinner party at our house just to try the stuff. I started out making a small amount of friends happy; now I make a lot of friends happy."

As for the business side of things, the ball started rolling when a friend of Morton's who worked for The Food Network went crazy for "the ginger stuff." She complained that she wasn't being invited over enough for dinner parties, and she couldn't get her fix. She told Morton he should go into business. But Josh wasn't convinced. He remembers telling her, "Who goes into the liquor business? That's insane."

But over time, he relented. During a weekend trip upstate, he spotted an article in *The River*

Reporter titled "Catskill Distillery Opening." Like all of us, Morton had come of age in a post-Prohibition New York utterly devoid of distilleries, and now he was witnessing the birth of one right in the vicinity of his own summer home. He mused, "'Okay, this is a weird coincidence, but maybe I could produce at the Catskill Distillery,' because I didn't know anything about it."

That didn't end up happening, but the fuse had been lit. Two years later, in February of 2013, Morton launched his new business venture, Proof Of Concept, LLC.

No stranger to the world of entrepreneurship, Morton today finds himself with two full-time businesses. Since they are both "nontraditional" businesses, Morton sets his own schedules. One night he may be up until three a.m.

working on his consulting; the next he may be up in the wee hours straining a new batch of Barrow's Intense (né "the ginger stuff"). But this breakneck pace won't last forever; it can't. He says: "There's time to do both, at least for the short term, until this is big enough that I don't have to do that anymore...because it's not really sustainable to do them both long-term."

Whatever Morton is doing, he's clearly doing something right. At his Industry City facility, plans are underway to move Barrow's into a much larger production space, around double the current size. The maceration phase would still be done in small batches, but every other aspect of production would be scaled up. While musing over what it takes to be a business owner, Morton smiled and recalled the words of his good friend and "Alcohol Fairy Godfather," Jack from Brooklyn: "When you're

an entrepreneur, it's all about overcoming insurmountable odds, and I like to surmount the fuck out of them." Jack helped Morton land his first three accounts and introduced him to The Bartenders Guild. Morton in turn helped Jack to get his business back up and running in the wake of Hurricane Sandy (see chapter 2).

Morton is not one to mythologize his production process. He is very up-front about how simple his product is to make. As he explains the process to us, we can almost picture his chef friend explaining the process to him all those years ago. "It's simply an infusion. You take your base spirit, and you put your ginger or whatever you're using in it, and you macerate that in the alcohol over a period of time. It's really simple. The two things that are key are how long you do it for, and what your ratio of sugar to water to alcohol and, in my case, ginger, is. It's not like it's a trade secret, in terms of how to do this sort of stuff."

Morton made the fascinating observation that people who were accustomed to having ginger as a part of their diets tended to notice the sugar more, whereas those unaccustomed to ginger in their diets really noticed the fresh ginger flavor.

And no wonder: Morton uses fresh, unpeeled ginger in his Barrow's, at the rate of a quarter of a pound per 750 milliliter bottle! There is a great deal of flavor in the ginger skin, Morton says, and the fact that there is no heat involved in the production really allows the character of that fresh ginger to shine through.

Another key to making Barrow's is that it is unfiltered. Morton strains his product using an ordinary—and not particularly fine—strainer. This method does leave some ginger particulates behind in the final product, but it also helps to preserve the flavor and, believe it or not, the color. During his experimental phase, Morton found that filtering stripped the flavor and the color right out, resulting in a less aesthetically pleasing clear liquid. Straining seemed to be the right approach.

Josh allows that he fully expects there to be differences in Barrow's flavor profile from batch to batch, and he likens that variability to the world of wine-making, where changes in flavor profile are expected from vintage to vintage. The changes in Barrow's flavor profile depend on the time of year the batch is made, because different gingers are available at different times. He prefers Brazilian ginger to Chinese ginger, but he can't get it year round. Then there's the matter of time. "The flavor changes as it ages in the bottle. So even if I produced a bottle a day, six months from now that flavor's going to drift slightly. It's a natural product. If I were using chemicals, then I could maintain a consistency, but I'm not. I get it within a range of a flavor, so that people know what they're expecting, but one batch may be a little more gingery than another, or have just a slightly different tone to it, but it will be in the same zone."

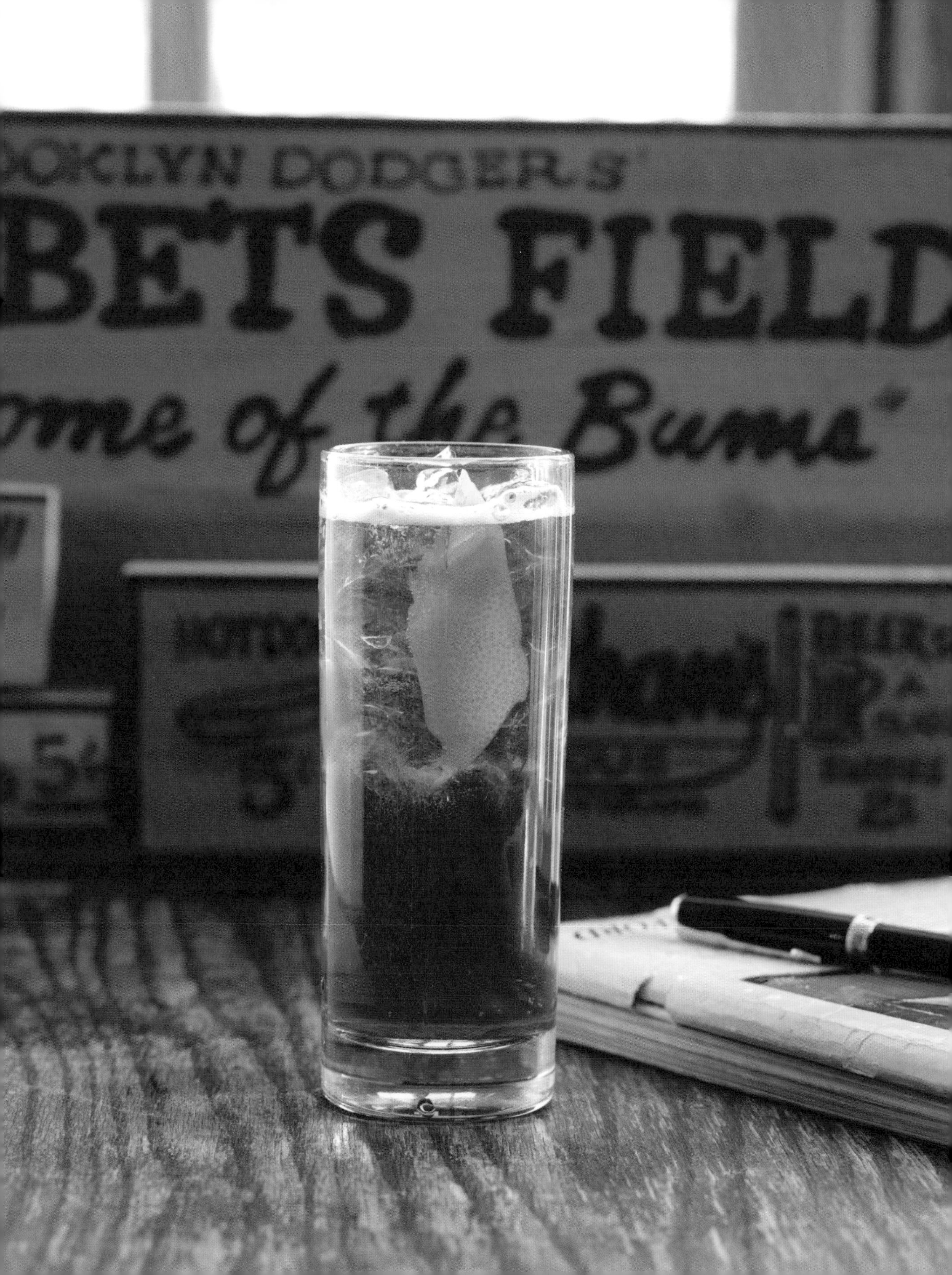
DODGERS'
of the Bums

EBBETS FIELD

2 oz Barrow's Intense
1/2 oz Fernet Branca
2 dashes Hella Aromatic bitters
4 oz club soda or seltzer (or to fill)

*Pour all ingredients into an
ice-filled collins glass and fill
with soda. Stir lightly to mix and
garnish with lemon twist.*

There's an old joke in Brooklyn:

Q: If a Brooklyn man finds himself in a room with Hitler, Stalin, and Walter O'Malley, but has only two bullets, what does he do?

A: Shoot O'Malley twice to make sure he's dead.

Pretty bitter stuff, eh? O'Malley was reviled borough-wide for his role in moving the Dodgers out west in 1958, thus robbing Brooklyn of its cultural identity and, some would say, its very soul. Of course, he was a facile target. Bitter Brooklynites could have just as easily blamed Robert Moses, but that would have required the sort of next-level thinking not common among hurting baseball fans.

This cocktail, the bitterest one we could think of, was designed in honor of those beleaguered fans, particularly John R. Quinn, whose memorabilia appears in this shot.

MOSCOW MULE

1 oz Industry Standard Vodka
1 oz Barrow's Intense
club soda
fresh lime wedge

*Build in a Moscow Mule
mug over ice.*

The Moscow Mule was invented in the 1940s by two industrious men who had a mutual concern in its popularity. One owned a ginger beer brand named Cock 'n Bull, and the other owned a vodka brand named Smirnoff. The Moscow Mule hit full fad status by the 1950s and became one of the key drinks that helped market vodka to the masses. Over time, it even got its own signature serving receptacle, a copper mug filled to the brim and garnished with limes. This reboot could just as easily be used to market Barrow's Intense to the masses in the 2010s. Bringing Josh's neighbors—Industry City Distilling—to the party honors the original recipe and adds the strength you'll need to kick a door down.

GINGER MULE

2 oz Brooklyn Gin

3/4 oz fresh lemon juice

1/2 oz Barrow's Intense

1/2 oz honey syrup (*see page 11*)

4 oz club soda (*or to fill*)

candied ginger (*available commercially or see below*)

raspberries

mint

Place the first four ingredients in a cocktail shaker, mostly filled with ice, and shake. Strain into rocks (or highball) glass and over ice, garnish lavishly with candied ginger, raspberries, and mint.

CANDIED GINGER

16 oz peeled ginger root

4 cups white granulated sugar

4 cups water

1/2 tsp non-iodized salt

Slice ginger as thin as possible (1/8 of an inch thick, max). Add the ginger slices to a stainless steel pot, add water to cover, and bring to a boil. Lower to simmer and simmer 10 minutes. Drain and repeat with more ginger. Combine 4 cups water and 4 cups sugar and add to the pot with the salt. Insert a candy thermometer into the pot and cook until syrup achieves 225°F. Remove from heat and let stand until completely cool, or drain ginger from syrup while still hot and coat slices with additional granulated sugar. Allow coated slices to air-dry on an oiled wire cooling rack overnight. Sugarcoated slices can be kept in an airtight container at room temperature for 2–3 months.

This is a relative of the Moscow Mule. No special mug is required, but you will want to make sure to seek out—or create—the right garnish. This one was designed by Damon Dyer of Brooklyn Gin. As he told us, "Fresh citrus and ginger have a certain affinity for each other. And Josh Morton makes some damned intense juice."

GINGER SHANDY

1 pint Six Point Crisp or other pilsner style beer

1 1/2 oz Barrow's Intense

dash or 2 Brooklyn Hemispherical Bitters sriracha bitters (*or just hot sauce, sriracha or otherwise*)

shichimi tōgarashi*

kosher salt

* a common Japanese spice mixture available in specialty stores or online.

On a plate, mix a teaspoon or so each of shichimi tōgarashi and salt. Prepare a pint glass by rubbing a lime around the outside of the glass and dip the rim in the tōgarashi/salt mixture. Fill the glass with ice, Barrow's Intense, and as much beer will fit. Top with bitters or hot sauce. Squeeze in a lime wedge.

We desperately wanted to have a beer cocktail in this book, and late in the process, two fell in our lap. One came from Lavender Lake, called the Damaged Soul. The other, Intensely Loaded, came from Josh Morton of Barrow's Intense. Unable to pick between the two, we started playing around during our photo shoot, and our food stylist and general consultant, Louise Leonard, devised this power-mix of the two.

CRAFT BEER IN BROOKLYN

The wonderful things going on with cocktails in Brooklyn are being mirrored in the beer world. We've both long been fans of Brooklyn Brewery (where Pete and his wife Susan had their wedding reception) and Sixpoint Craft Ales, and Brooklyn remains on the cutting edge of what's cool in beer. From new brewers like The Other Half and Grimm Ales, to great bottle shops/bars like Bierkraft and St. Gambrinus, to great pubs like The Owl Farm and The Gate, to innovative beer palaces like Spuyten Duyvil and Tørst, there is a whole lotta shakin' going on when it comes to the Brooklyn beer scene. Keep that in mind when you need to cleanse your palate before the next cocktail.

TRANCE OF THE HARPSICHORD

1 oz Barrow's Intense
2 oz Busnel Calvados
1/3 oz of Absinthe Ordinaire
1/4 oz of fresh squeezed lemon juice
thinly sliced fresh ginger*

Pour the first four ingredients in a shaker mostly filled with ice. Stir until your fingers are freezing. Strain into a chilled cocktail glass. Garnish with freshly sliced ginger that's been soaked in a little more of the Busnel Calvados.

*To prep the garnish, slice a long peel of skinless ginger that's been soaked in Busnel Calvados for at least an hour.

Matty McDermott of The Good Fork is a tremendous supporter of the local spirits scene, including Barrow's Intense. This is a Sidecar-inspired riff he came up with one night when he got the idea to pair ginger liqueur and absinthe—a combination that works surprisingly well.

"It took owner/manager Ben Schneider and me about eight hours of deliberating to come up with this one. I'm a musician and I wanted to call the drink The Ginger Baker—he's an amazing drummer and I'm a huge fan, but Ben wasn't into that. We went back and forth about names all night. I wanted something to convey the more expressionist, absurd, artistic side to the absinthe and it just came to me: The Trance of the Harpsichord."

Brooklyn Gin's founders, Joe Santos and Emil Jättne, are both spirit industry veterans who met while working for one of the biggest spirits companies on the planet. They bonded over their desire to do things differently from the corporate status quo. So, like others on the local distilling scene, they left lucrative, secure, and successful careers behind them to pursue their dream of opening a business making craft spirits. ▶

Said Santos, "I had some great mentors in the industry who are master distillers and basically made everything under the sun, and I really fell in love with the whole process. Even though I started out in marketing, I ended up really loving the whole concept of making something from scratch, from just an idea in your head to being an actual spirit on the shelf."

For Santos to really engage his creative side, he knew he'd have to start an independent business. "There were a lot of politics in the corporate world, and I got sick of the handcuffs that were put on my creativity to make new stuff."

Enter Emil Jättne. "When Emil and I met, he was going through some similar frustrations, and we started talking and realized that we both had a real aspiration to do something on our own, to do it our way, without compromises, and not having accountants, engineers, and lawyers constantly telling us 'No.'"

But why Brooklyn and why gin? Jättne said, "Brooklyn has so many like-minded neighbors. They come to Brooklyn because they share similar values and aspirations. They have big ideas. They want to contribute. They want to build something. We chose this name because it meant something to us. It articulated our values and aspirations."

"Emil and I have talked about what Brooklyn means in Sweden, where he's from," said Santos. "There's the artisanal side that we were talking about, the whole do-it-yourself movement. That idea of ownership was instilled in Emil from an early age. His father owned a lumberyard and his mother owned a flower shop in

Sweden. Now, even abroad, Brooklyn is seen as the epicenter of the craftsman movement. I think that's why Emil wanted to raise his two children here in Park Slope."

While they maintain a lofted workspace in the neighborhood of Gowanus, they distill their gin in the Hudson Valley, where they have partnered with one of the oldest craft distilleries in the state, the beautiful Warwick Valley Winery and Distillery. There, surrounded by forests and fruit orchards, they make 300 bottles of gin at a time over three days, before shipping it down to Brooklyn.

Some have questioned the assertion that a "Brooklyn" gin could be made in Upstate New York (hey, to a true Brooklynite, anything north of Fordham University's Rose Hill campus is considered Upstate). But it seems

very much in keeping with the Brooklyn ethos that Santos and Jättne chose to get the exact product they wanted made by whatever means necessary. "We wanted to make our gin our way with no compromises," Jättne said. "Warwick Distillery has one of the best stills for the gin we wanted to make, and the Hudson Valley is great. We love living and working in Brooklyn, but we also enjoy getting our hands dirty at the distillery. Not a bad place to make gin."

Why did Santos and Jättne decide to focus on gin and gin alone? "Gin was actually one of the first ideas that popped up. We love gin. Emil's go-to drink has always been the Gin and Tonic, and I really got into gin through the cocktail scene. So I was doing more of the classic cocktails, and when we started really looking at what we were both passionate about, gin was the clear winner. We wanted to pick a spirit where we could make an impact in the category and in the industry."

Santos and Jättne have done just that. Brooklyn Gin performed well in several informal tastings conducted with industry types. It is currently available all over the Northeast and in California, plus internationally in Spain, Sweden, Singapore, Australia, and New Zealand.

"There are a couple of things we do that really stand out versus all other gins," Santos explained. "One is the way we use citrus peels. Citrus is not an uncommon element in gin. Typically what's used is dried or frozen peels. We're one of the few gins in the world that use 100 percent fresh citrus peels. So when we were starting out, I was running down to Red Hook, buying all my stuff over at Fairway, cutting and peeling in my kitchen."

> ## "Brooklyn has so many like-minded neighbors. They come to Brooklyn because they share similar values and aspirations. They have big ideas. They want to contribute. They want to build something."
>
> **—EMIL JÄTTNE**

Jättne explained that when they were searching for a juniper to use, they turned to a friend of his who was once a chef at the famed Copenhagen restaurant Noma. He told them to try Albanian juniper for its extremely aromatic properties and naturally occurring fresh oils. Santos continued: "And then the other kind of unique thing we do is we crack open our juniper berries. It's labor intensive. We don't have a mill, so we have to lay them out on a baking sheet and we just crack 'em open with a frying pan." Jättne added, "Cracking the juniper releases all the sweet, spicy flavors within, which works great with Brooklyn Gin's fresh citrus flavors." "We tested making the gin without these particular steps, by using dried or frozen citrus, by using un-cracked juniper, and it does make a difference," added Santos.

Another interesting element to the Brooklyn Gin recipe is the inclusion of cacao nibs. We thought the cacao sounded interesting, but we were intrigued to learn that it isn't technically a part of the intended flavor profile. "The cacao's actually more for function than it is for flavor," said Santos. "Because we use fresh citrus peels, we get a lot of acidity, and we use the cacao nibs to balance things out. It would be very rare that I'll find someone who actually picks up the chocolate note—we don't really use enough of it to give it a strong chocolate note, but it is unique."

Smart spirit-makers know that their product is powerless without some good ideas about how to use it, so many of the local makers have created relationships with leading figures in the cocktail world to help get the word out about their products. Brooklyn Gin has gone a step further. The New York Brand Ambassador, Damon Dyer, is a veteran of the New York City cocktail scene, having done stints at Jack the Horse Tavern, Flatiron Lounge, Clover Club, and Lani Kai. Santos said, "Damon brings a skillset to us that neither Emil nor I had. He's a great asset not only as far as using his connections as a salesman, but also through educating people on how to use their gin well."

THE BOTTLE

The bottle for Brooklyn Gin stands out among all of the local craft spirits as a real work of art. Its aquamarine color is reminiscent of Depression glass—light, colored containers that were given out free by companies during the Great Depression—and evocative of a time in New York City when the finest gins were made in bath tubs. The shape is inspired by classic Art Deco architecture in Brooklyn like the Williamsburgh Savings Bank Tower in Fort Greene. The round emblem is the bottle's most notable feature. According to Santos, he and Jattne chose copper to allude to their copper pot still, and the shape is meant to evoke the urban elegance of a New York City manhole cover.

What's in the bottle is even more special.

SEASONAL BROOKLYN GIN AND TONICS

2 oz Brooklyn Gin
4 oz tonic water

Pour the gin in a collins glass filled with ice and top with tonic water. Garnish with fresh, seasonal ingredients as suggested here.

"To every thing there is a season, and a time to every purpose under heaven."
—Ecclesiastes 3:1

It's impossible to know now if the accompanying Biblical passage referred to Gin and Tonics, but we suspect it might have. Since almost every gin has to fit into its most consumed cocktail, we asked Damon Dyer to give some options that work particularly well with Brooklyn Gin year-round. The following were his suggestions, and you should feel free to get creative and make some of your own (sugar snap peas in spring and pomegranate seeds in winter?):

Seasonal Brooklyn Gin and Tonics—all served in a collins glass with a clear ice spear and great tonic water. Grab a straw and turn, turn, turn.

SPRING

Garnished with the peels of entire fresh navel orange, lemon, and lime. The essential citrus oils from the peels slowly release additional flavors into the cocktail, and the tonic's effervescence brings the fragrant citrus peel aromatics to the surface, highlighting the fresh citrus in the gin and evoking the fresh smells of spring.

SUMMER

Garnished with fresh cut lime, lemon, ruby grapefruit, and kumquat. Bright and vibrant, the bitter grapefruit is a refreshing palate cleanser in the dog days of summer.

WINTER

Garnished with fresh Meyer lemon and blood orange. Using winter citrus fruits at the peak of their season, this is a terrific cold weather cocktail. The intensely sweet perfume of the fruit will evoke warmer days to come, transporting you beyond the cold winter night and into your next drink.

AUTUMN

Garnished with fresh lemon, juniper berries, and a rosemary sprig in a way that underlines the spice and complexity of the gin. The spice and herb elements are the perfect complement to the fruits of the harvest you'll find on fall tables.

BROOKLYN HEMISPHERICAL BITTERS

Many people start their own businesses, but in the case of Brooklyn Hemispherical Bitters, according to Mark Buettler, "the business started us." Buettler was tending bar at the renowned and pioneering Brooklyn restaurant Dressler when he began making bitters as a matter of economic practicality. The Bitter Truth had the only celery bitters on the market, and they had to be ordered by mail from Germany at no small cost.

Mark began researching recipes and experimenting with different flavors. He began handing out free samples to his bartender friends. One day Buettler's former business partner sold the product to a bar in Australia (hence the "Hemispherical" of the name)—a rather delicious irony for a company that began because Buettler didn't want to buy bitters from Germany! "One thing led to another," he said, "and then people started asking for them, and I decided I had to start a business."

Today, Buettler is the sole proprietor of Brooklyn Hemispherical Bitters. Mark makes four main flavors of bitters and several seasonal offerings. The four main flavors are Meyer Lemon, Rhubarb, Black Mission Fig, and Sriracha. Yes, you read that last one right. Mark explained, "I always had a weird dream of doing sriracha bitters. I just have a fascination and love for sriracha. I didn't even know how I wanted to use them cocktail-wise. I just wanted them to exist." (We had one idea that we presented back on page 242.)

Mark wants to concentrate on unique, single-flavor expressions. We asked for a rundown on his other three bottlings. "We wanted a citrus, but we wanted something unique. So we worked with the Meyer lemon; it ended up being pretty awesome. Black Mission Fig is also a really great flavor. And because black mission figs are in season spring and fall, we pretty much have them year-round. Rhubarb bitters are available elsewhere, but we really liked how ours turned out. With the limited season of rhubarb, there's always a chance we'll run out, but that's just the way it is."

THE BEE'S KNEES

2 oz Brooklyn Gin
3/4 oz wildflower honey syrup (*see page 11*)
1/2 oz simple syrup (*see page 11*)

Combine all ingredients besides the garnish in a shaker mostly filled with ice. Shake and strain into a rocks glass filled with ice. Garnish with a lemon wheel.

This Prohibition-era cocktail can be pronounced two ways: the obvious one, "The Bee's Knees," or more esoterically, "The Business."

The irony of this cocktail is that while it was designed to mask impurities in Prohibition-era spirits, it actually does an excellent job featuring its base spirits due to the spareness and quality of the other ingredients. As a result, it's become very popular on the Brooklyn cocktail scene for various gins and even using Van Brunt Moonshine (see page 99). Or as Joe Santos puts it, "The Bee's Knees is another recipe we love, because the cocktail itself is so simple it allows you to taste the complexity of the gin."

TUX TOO

2 oz Brooklyn Gin

3/4 oz Dolin dry vermouth

1/4 oz maraschino liqueur

1/4 oz Tenneyson Absinthe (*if substituting a different absinthe, just rinse the glass*)

2 dashes orange bitters

Stir well and strain into a chilled cocktail glass or coupe that has been prepped with a lemon twist. Discard the twist or deposit it in your glass as you wish. Note that the Craddock version of the recipe suggests a cocktail cherry.

On a recent trip to Maison Premiere, we somewhat obnoxiously ignored the carefully constructed drinks menu and simply asked for a cocktail that featured "one of the local gins and Tenneyson absinthe." Our guy thought for a minute, then reached back into the classic Harry Craddock tome *The Savoy Cocktail Book* and brought us a Martini variation that he called a Tux Two—it's known as Tuxedo Cocktail (No. 2) in the Craddock book.

The drink was perfect, with the different liquors and liqueurs each adding a level of flavor, while the citrus and juniper notes of the Brooklyn Gin were still noticeably present in the background. Here's how to recreate the drink at home.

MAXWELL'S RETURN

2 oz Brooklyn Gin

1 oz fresh pineapple juice

1/2 oz fresh lime juice

1/2 oz simple syrup (*see page 11*)

1/2 oz green Chartreuse

Muddle one rosemary sprig in the bottom of a cocktail shaker with the simple syrup. Add ice and the other ingredients, shake, and strain into a rocks glass filled with ice.

This is a cocktail that proves how close the cocktail community is in a borough of four million people. Damon Dyer made this drink in honor of a former protégé of his at Jack the Horse Tavern in Brooklyn Heights. He said, "This cocktail is named after my old friend Maxwell Britten for when he was convalescing from a car accident and needed prescription-strength booze to recover." What a friend. Britten is now beverage manager at one of the most esteemed cocktail bars in America, Brooklyn's own Maison Premiere.

MAISON PREMIERE

(Williamsburg)

Walking into Maison Premiere in Williamsburg is a lot like walking onto a movie set. There are beautiful people, neatly chosen props, carefully costumed extras. It's a mannered place that people tend to either love or hate. Count us in the former category. For one thing, Maison presents a one-of-a-kind experience. The décor is ostentatious and amazing: the New Orleans-inspired horseshoe bar, the absinthe fountain, even the vintage toilets. And the details only add to the experience: the copper oyster tubs, the yellow pine floorboards, and the 19th-century fashions sported by the guys behind the bar. OK, so the arm garters might be a bit much, but they also definitely speak to a certain seriousness and sense of craft. Most important of all, we have yet to receive a poorly made drink in Maison Premiere. All that attention to detail translates into pleasure in the glass.

Started by Krystof Zizka and Joshua Boissy in 2009, Maison has vaulted to the top of various "Best Bar" lists in Brooklyn and nationally. In her profile of Maison Premiere in a 2013 issue of *Edible Brooklyn*, Rachel Wharton had a wonderful line about how head bartender Maxwell Britten and his colleagues "make time machines by the glass."

FLOYD BENNETT COCKTAIL

2 oz Brooklyn Gin
3/4 oz fresh lemon juice
3/8 oz crème de violette
3/8 oz fresh lavender syrup*
orange zest
lemon wheel

Place all ingredients except the lemon wheel in a cocktail shaker with ice. Shake hard and double strain into a cocktail glass or coupe. Garnish with the lemon wheel.

*To make the fresh lavender syrup, take 2 tablespoons of lavender stems and flowers, combine with 1/2 quart sugar and 1/2 quart water, and simmer until it just starts to boil. Remove from heat, let steep for 10 minutes, then let cool and strain into a container for storage.

Johnny DePiper of Featherweight in Williamsburg has his own take on the Aviation, which we've named for the book in honor of legendary Brooklyn-based aviator Floyd Bennett. Featherweight is a cool spot in Williamsburg, sort of a speakeasy connected to the restaurant Sweet Science. It's a dark, small, secluded spot, perfect for a pre- or post-dinner drink, or secret rendezvous.

THE AVIATION

God bless the cocktail geeks. Some of the research that's been done over the last decade has really made drinking a more fun and rewarding experience. Take for example the re-introduction of crème de violette to the marketplace.

When Pete first learned to make an Aviation from Dale DeGroff's *Craft of the Cocktail*, it was a drink that consisted of 2 ounces of gin, ¼ ounce of maraschino liqueur, and 1/2 an ounce of lemon juice. The name made no sense, but that was OK. We still have no idea why the Alaska is called an Alaska and it doesn't bother us in the slightest.

But at some point, the truth was revealed. When the Aviation recipe was listed in *The Savoy Cocktail Book*, an ingredient was left out, a key ingredient, if you want the names of your drinks to make sense. Add a scant 1/4 ounce of crème de violette and the name becomes as clear as the nose on your face—the drink takes on a gorgeous sky-blue hue).

BROOKLYN CROSSING

2 oz Brooklyn Gin
1/2 oz dry vermouth
one tsp full of brine
various pickles for garnish*

*(Gabrielle changes up the pickles
seasonally, but the mix typically includes
a combination of cucumbers, green beans,
turnips, olives, or cocktail onions)*

*Put all the ingredients in a shaker filled with
ice. Stir vigorously and strain into a chilled
cocktail glass. Garnish with skewered pickles.*

Nothing says artisanal Brooklyn like locally made pickles. We've encountered four artisan pickle-makers at least partially based in Brooklyn while working on this book: McClure's, Rick's Picks, Brooklyn Brine, and The Pickle Guys. Pickling has become so ubiquitous that cocktail historian and Brooklynite David Wondrich has named the adjoining neighborhoods of Carroll Gardens, Cobble Hill, Boerum Hill, Gowanus, and Park Slope the "Pickle Belt."

For this briny take on a classic Martini, Gabrielle Delaney of The Castello Plan reached for The Pickle Guys. Started by Alan Kaufman on the Lower East Side, Pickle Guys' second location has opened in Coney Island, which is good for us because who wants to do a book about Brooklyn in which Coney Island doesn't at least get a mention?

I WILL FIND YOU

2 oz Brooklyn Gin
1/2 oz Hidalgo Fino Sherry
1 oz freshly squeezed
grapefruit juice
3/4 oz honey syrup (*see page 11*)
dash of Angostora aromatic
bitters

*Shake with ice and 2 sage
leaves. Fine-strain and
serve up in a coupe glass.*

This drink was invented by Ian Hardie, the senior bartender at Huckleberry Bar. We asked bartender Jo Ann Guerrero, who spent an afternoon with us making cocktails and talking local spirits, about why the drink works so well with Brooklyn Gin. Guerrero likened the Brooklyn Gin to an actress: "She is the leading lady who doesn't need to be the center of attention. Sure, she can perform very beautifully on her own, but she really shows her range when paired with a few choice ingredients like honey and sherry, and the sage really rounds everything out with its earthy savory characteristics."

Bruce Arthur, former bartender at Huckleberry, told us the story of the cocktail's name: a reference to *The Last of the Mohicans*. "At a planning meeting in 2013, I suggested that this then-unnamed drink tasted like the air behind the waterfall in the famous scene in that movie and suggested that it should be called Behind the Waterfall. 'No,' interjected bartender Meaghan Montagano. 'You should call it I Will Find You, after the iconic line in that scene.'

"Meaghan managed to capture the urgency and romance of that scene and apply it to the drink. We all agreed and understood. Probably no one else will unless they're reading this."

PICKLED RAMP GIBSON

2 1/2 oz Brooklyn Gin
1/2 oz dry vermouth
1 pickled ramp
1/4 tsp pickled ramp brine

In a cocktail shaker filled with ice, add the gin, vermouth, and brine. Stir vigorously and strain into a cocktail glass, preferably chilled. Garnish with a pickled ramp.

PICKLED RAMPS

1 lb ramps
8 oz white wine vinegar
1 2/3 tbsp white granulated sugar
1 1/2 tsp non-iodized salt
2–4 juniper berries
2–4 black peppercorns

Remove the greens from the ramps for another use. Trim the bulbs. Wash thoroughly and blot dry. Simmer vinegar, juniper, salt, and sugar in a saucepan. Add the ramps and simmer on low heat 15 minutes. Remove from heat and cool in pan. When cool, transfer pickled ramps to a Mason jar or other sterilized, airtight container. Let combine for at least 30 minutes before serving. Ramps will last 3–4 months in the refrigerator.

Francine Stephens, of her namesake restaurant Franny's, has a unique philosophy towards cocktails.

"If you start with a classic cocktail and then take it in a slightly different direction that is true to who you are, that just makes sense. And that's what we wanted to do with our Gibson. A Gibson is a great drink, except for the fact that I hate those little commercial cocktail onions—so we took that out of the equation and replaced it with a delicious pickled ramp. It's seasonal; it's simple; it's straightforward. The ramp looks beautiful in the glass and it adds a sweetness to the drink that works well with the brininess."

This drink has appeared on the Franny's menu using various gins, including New York Distilling's Dorothy Parker. But we feel that Brooklyn Gin, with its prominent citrus notes, is a natural fit for this great spring/summer cocktail. "It's bright and that creates a really nice juxtaposition with the pickling liquid," Stephens told us. "All that briny character is there, but the first thing I'm getting is the brightness. And for me, that works really well."

A NIGHT IN TUNISIA

2 oz Brooklyn Gin
3/4 oz fresh lemon juice
3/4 oz kumquat syrup
1/4 oz honey syrup (*see page 11*)
2 dashes Hella Bitter Citrus bitters

Combine the ingredients, shake, and strain into a rocks glass over crushed ice. Garnish with fresh kumquat speared to a lemon wheel.

KUMQUAT SYRUP

24 kumquats
1 cup water
1 1/4 cup sugar
1 oz Brooklyn Gin

Lightly zest kumquats, saving the zest and the fruit. Bring the water to just under a boil. Next, cut the kumquats in half and combine them all with the hot water and the sugar. Remove from heat, add Brooklyn Gin to fortify, stir, and cover. Let the mixture sit for eight hours. Strain and bottle.

This is another great submission from Damon Dyer that builds on the music theme that resonates throughout this book. Damon explained, "I take great inspiration from jazz tunes while naming cocktails, and there are few songs greater (to my tin ear, at least) than 'A Night in Tunisia.' Dizzy Gillespie didn't write a ton of songs, but he wrote (and performed) some quality shit."

As for the drink: "Fresh kumquats are key to Brooklyn Gin's distillation and are our most identifiable citrus botanical. The pulp may be tart, but the peels have a round sweetness that is vibrant and damned delicious. Without the humble kumquat, Brooklyn Gin would not have the citrus complexity that defines it."

INDEX

Brooklyn Spirits: Craft Distilling and Cocktails from the World's Hippest Borough

Photographs and text © 2014 Peter Thomas Fornatale and Chris Wertz

Recipe on page 105 reprinted with permission from
Bitters: A Spirited History of a Classic Cure-All by Brad Thomas Parsons,
published in 2011 by Ten Speed Press, Berkeley, California.

Published in the United States by powerHouse Books,
a division of powerHouse Cultural Entertainment, Inc.
37 Main Street, Brooklyn, NY 11201-1021
telephone 212.604.9074, fax 212.366.5247
e-mail: info@powerHouseBooks.com
website: www.powerHouseBooks.com

First edition, 2014
Library of Congress Control Number: 2014942948
ISBN 978-1-57687-705-0

Design: Eric Skillman
Photography: Max Kelly
Food Stylist: Louise Leonard
Endpaper illustrations: Josh Cochran
Production Assistant: Lindsay Berenson
Printed by Toppan Leefung Printing Ltd

10 9 8 7 6 5 4 3 2 1

Printed and bound in China

The authors wish to formally acknowledge all the men and women
whose names appear in this book. We couldn't have done it without you.

Many thanks also to Craig Cohen, Will Luckman, Stephen Shellenberger, and Teresa Genaro for their help.
And a special thanks to Thomas E. Harkins for his writing help and recipe testing ability.

www.brooklyn-spirits.com